# Floaters and Sinkers

## Authors

Betty Cordel
Dave Youngs
Arthur Wiebe
David Lile

Kathleen Landon
Mike McKibban
Walt Laidlaw
Greg West

**Desktop Publishing**
Tanya Adams

**Illustrator**
Reneé Mason

**Editor**
Betty Cordel

# Floaters and Sinkers

This book contains materials developed by the AIMS Education Foundation. **AIMS** (**A**ctivities **I**ntegrating **M**athematics and **S**cience) began in 1981 with a grant from the National Science Foundation. The non-profit AIMS Education Foundation publishes hands-on instructional materials (books and the monthly magazine) that integrate curricular disciplines such as mathematics, science, language arts, and social studies. The Foundation sponsors a national program of professional development through which educators may gain both an understanding of the AIMS philosophy and expertise in teaching by integrated, hands-on methods.

ISBN **1-881431-58-4**

Printed in the United States of America

# Floaters and Sinkers

## Table of Contects

# Density Demystified

**Density** is a measure of the "compactness" of a material. It is the ratio of mass to volume for any material measured in grams per cubic centimeter and tells how much matter is packed into a given space. Density is not a simple comparison of the "heaviness" or "lightness" of materials. It is instead, a comparison of the "heaviness" or "lightness" of the same volume (mass per unit volume) of materials. A steel pin has the same density as a steel beam. The beam is thousands of times more massive than the pin, but it also takes up thousands of times

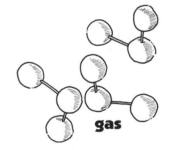

more space, so the ratio of mass to volume (density) remains the same for both (about 7.8 g/cm³).

The density of a small pebble is greater than the density of a huge redwood tree even though the tree is much larger. The density of a kilogram of feathers is much less than the density of a kilogram of gold even though they have the same mass. The mistake most people make in thinking about density is to consider only the size or mass instead of both of them together. **When dealing with density, mass and volume always go hand in hand.**

The density of materials is determined by the masses of the atoms in the material and the amount of space between the atoms. Gases have a low density not only because the atoms making up the gases have a small mass, but also because there is a large amount of space between the atoms. The heavy metals like gold, lead, and uranium are very dense because the atoms they are composed of are massive and spaced closely together.

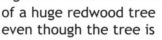

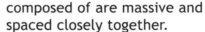

Water has a density of one gram per cubic centimeter at four degrees Celsius and is the standard for comparing the density of materials. Materials with a density greater than one gram per cubic centimeter are denser than water and will sink in water; materials with a density that is less will float in water. Lead has a density of 11.3 g/cm³ which tells us it is more than eleven times as dense as water. This means that a cup of lead would have eleven times more mass than a cup of water. It also means that a hundred grams of lead would have eleven times less volume than a hundred grams of water (the atoms in lead are much more closely packed together so they take up less space).

This brief explanation of density shows the difficulty in dealing with certain scientific concepts that are rather abstract and not easily understood. This difficulty is compounded because the concepts are often introduced in science books by a brief definition, explanation, or formula that is very abstract to students and hard to internalize. Students learn to memorize definitions and formulas to pass science tests, but do not necessarily develop the mental structures for really understanding the concepts that are presented in the science textbooks.

Density is one of the concepts not readily understood by students because it is usually introduced to them in the form of a mathematical formula: Density = mass/volume. While this formula accurately describes how to calculate density, it doesn't necessarily help students understand the concept of density.

Paul Hewitt, an outstanding physics teacher and the author of *Conceptual Physics,* contends that it is important to build concepts in physics before dealing with the calculations. This **"concepts before calculation"** approach can be expanded to include **"concepts before definitions"** and is a vital approach to use if students are to develop mental structures for understanding density or any other scientific concept.

To help students develop the concept of density, we must look at how they learn. Piaget would have students start at the concrete level in order to understand a concept at the abstract level. By looking at learning theory, we see that students need to be given concrete experiences with density to help them develop the concept behind it before they deal with the abstract density definition and formula.

A concrete starting point to help students learn the concept of density might be the simple investigation and discussion that follows. This activity can be initiated by asking students whether steel or wood is heavier. The immediate response would most likely be steel. The teacher can then give each student a paper clip and have them compare its mass to a wooden ruler or pencil from their desk. The teacher can then tease the students and say that wood must be heavier than steel since the ruler, which is made of wood, is heavier than the paper clip, which is made of steel.

The discussion following this investigation should make it quite clear to the students that the question they were asked is very misleading. They should also come to the conclusion that mass can only be used to compare specific objects: a baseball is heavier than a marble. But mass alone cannot be used to compare   different types of materials. It is erroneous to say that glass is lighter than lead without taking volume into consideration (the volume of glass may be much greater than the volume of lead). The students can then be asked how they could use mass to compare types of materials. The dilemma of how to compare the mass of types of materials and the ensuing discussion should help students "discover" the concept of density where they compare the masses of the same volumes of different materials. Having "discovered" a way to compare various materials, the students are ready for the formal definition of density and its formula.

The next logical step in the teaching sequence is to do investigations with the students where they compare the densities of various materials rather than to jump into abstract work problems to apply the density formula. One such investigation might be to use identical cups or containers and fill them to the top with various liquids. The cups would all contain the same volume (amount) of liquids but the masses of the materials would most likely be different. Using a balance, the cups could then be arranged according to the mass of the materials they contain. This investigation reinforces the concept of density by comparing the "heaviness" (mass) of identical "amounts" (volumes) of various substances. There is not even a need to attach numbers to the investigation at this point. The materials can simply be ordered sequentially; material A is most dense, material C is least dense, and material B has a density between the densities of materials A and C.

2

After the students have a well-developed mental structure of the concept of density and can calculate it, they might do two types of investigations. In the first, they find equal volumes of various materials and use a balance to find and compare their different masses.

In the second, they use a balance to find equal masses of various materials and then use a graduated cylinder to compare their volumes (as long as the materials being compared can be "poured" into a graduated cylinder). At this point, they should be able to see that the least dense material would have the greatest volume and that the most dense would have the smallest volume.

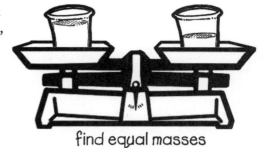

find equal masses

Having had many hands-on experiences, the concept of density should be more real to the students and they should have no problem doing density problems and applying the density formula because they have developed the proper mental structures to understand density. They should also be able to reflect this understanding on a paper-and-pencil test which covers density, although a better test would be to have them find the mass and volume of some material and then calculate its density in a real-world test.

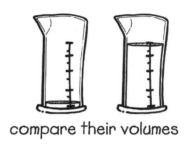

compare their volumes

# Floating and Sinking

Floating and sinking are such common phenomena we often don't give them much thought. Why do some things float while others sink? How can ships made of steel float? How do submarines both float and sink? Why do hot air balloons float in the air? In order to accurately answer these questions, we must take a closer look at three scientific concepts and the principle that ties them together.

**Gravity** is a mutual attraction between bodies that is proportional to the masses of the bodies and the distance between them. Therefore, the Earth attracts us and we attract the Earth. Everything within the Earth's gravitational field is attracted towards the center of this field in the Earth's core. Only the ground prevents most things from being pulled to the Earth's center.

*Sir Isaac Newton's apple really wanted to fall all the way to the center of the Earth, but was prevented from doing so by the solid ground it encountered under the tree.*

**Density** tells us about the compactness of a material and is determined by the masses of the atoms that make up the material and how close together those atoms are. In scientific terms, density is a ratio of mass per unit volume and is usually measured in grams per cubic centimeter. When comparing the densities of two materials, both volume and mass must be taken into account. For example, iron is denser than aluminum, therefore one cubic centimeter of iron has more mass than one cubic centimeter of aluminum. Likewise, mercury is more dense than oil, so one hundred grams of mercury takes up much less volume than one hundred grams of oil. Water is a standard for measuring density. It has a density of one gram per cubic centimeter at four degrees Celsius.

*The mercury and the oil have the same mass, but different volumes. Since the oil has more volume, its density is less than that of the mercury.*

**Buoyancy** is an upwards force acting on all objects in fluids, whether they are floating or submerged. It is caused by pressure differences within the fluids. Pressure in fluids increases with depth, thus the pressure on the bottom of an object is greater than the pressure on its top, causing a net upwards force, called buoyancy. This force acts against gravity and explains why heavy objects are easier to lift when they are in water.

*Pressure in a liquid increases with depth. The greater pressure on the bottom of the rock produces an upward buoyant force.*

Now let's see how gravity, density, and buoyancy are related to floating and sinking by looking at what happens to things in water. As mentioned above, gravity affects everything on Earth. This includes bodies of water and the things in them. Gravity tries to pull the water and the things in it down toward the center of the Earth. Only the solid surfaces beneath bodies of water prevent this from happening to the water and objects that sink in it. Some materials, however, such as wood, Styrofoam, and oil, float in water and overcome gravity's downward pull. These materials are less dense than water and are buoyed up by the water. This upward buoyant force of the denser water on the less dense materials overcomes the force of gravity which would otherwise make them sink to the bottom.

Materials that are denser than water are also buoyed up by the water, but not enough to overcome the force of gravity. The pull of gravity on materials denser than water is greater than the buoyancy imparted by the water, causing them to sink to the bottom. Thus, one factor in determining whether things will float or sink, is their density in relation to water. Any material or object that is less dense than water will overcome the force of gravity and float in water.

The other key factor in determining whether things float or sink is buoyancy. **Archimedes' principle**, which was first formulated in the third century B.C. by the Greek mathematician and scientist Archimedes, explains buoyancy's role. It states that a body completely or partially submerged in a fluid (a fluid is either a liquid or a gas) is buoyed up by a force equal to the weight of the fluid it displaces. If the weight of the fluid displaced by an object is equal to the weight of the object, the object will float. If the weight of the fluid displaced is less than the weight of the object, the object will sink.

Archimedes' principle explains why a ship made of steel, or other materials denser than water, floats. A block of steel is about eight times denser than water. When placed in water, the block of steel doesn't displace enough water to equal its weight, so it sinks. If this same block of steel is formed into a ship's hull that displaces eight times as much water as the original block did, it will float. Thus, the shape of objects is an important factor in floating or sinking.

A submarine is unique in that it can adjust its density to either float or sink. When its ballast tanks are filled with air, its density (mass/volume) is less than that of water and it floats. When air is let out of the tanks, they fill with water and the submarine sinks. Replacing the air in the ballast tanks with water does not change the volume of the submarine, but it does increase its mass and therefore its density. The submarine can also adjust its density to equal that of the surrounding water and sit suspended beneath the surface. When the water in the ballast tanks is forced out by compressed air, the submarine's density is again changed and it floats to the surface.

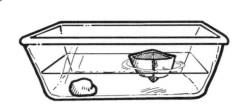

*A lump of clay will sink in water because it is more dense than water. If the clay is molded into the shape of a boat, however, it displaces an amount of water equal to its weight and floats, demonstrating Archimedes' principle.*

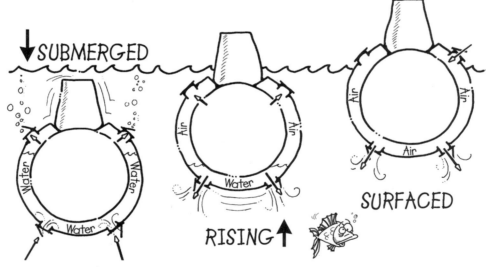

A hot air balloon operates much like a submarine and demonstrates the same principles. The major difference is that the hot air balloon operates in a sea of air. Since Archimedes' principle applies to gases as well as liquids, we know that objects on the surface of the Earth are buoyed up by the sea of air surrounding them. This buoyancy, which is equal to the weight of the air displaced, is not enough to overcome gravity for most things. However, if an object displaces a volume of air that weighs more than it does (like a helium-filled balloon), it will float in the air. An object which displaces a volume of air that weighs less than it does (like a normal balloon), will sink in the air. Hot air balloons, like submarines, can vary their density. They do this by changing the temperature of the air they contain. The density of air is inversely proportional to its temperature; the hotter the air, the less dense. Thus, a hot air balloon can vary its density to either float or sink in the air.

It should now be obvious that floating and sinking are not simple phenomena. An understanding of gravity, density, buoyancy, and Archimedes' principle are needed to really understand floating and sinking.

# WHICH WAY IS UP?

**Topic**
Volume

**Key Question**
What is the length, width, and height of each box?

**Learning Goal**
Students will measure the lengths, widths, and heights of various boxes to determine their volumes.

**Guiding Documents**
*Project 2061 Benchmark*
• Calculate the circumferences and areas of rectangles, triangles, and circles, and the volumes of rectangular solids.

*NCTM Standards 2000\**
• Select and apply techniques and tools to accurately find length, area, volume, and angle measures to appropriate levels of precision
• Understand relationships among the angles, side lengths, perimeters, areas, and volumes of similar objects
• Understand the meaning and effects of arithmetic operations with fractions, decimals, and integers

**Math**
Measurement
    linear
Calculations
Use formulae
    volume

**Integrated Processes**
Observing
Classifying
Collecting and recording data
Generalizing
Applying

**Materials**
Eight boxes or blocks with at least one pair of the same dimensions
3 colors of construction paper
Colored pencils, optional (see *Management)*

**Background Information**
Volume is the capacity of a three-dimensional object. In this activity students will figure volume by multiplying length times width times height. The result should be recorded in cubic units. Students are often confused with which measurements are considered as length, width, and height. This investigation is to help them visualize these concepts and their application. The longer base dimension of a box is called the *length;* the shorter base dimension is called the *width.* The *height* is the side perpendicular to the base.

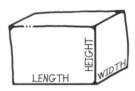

**Management**
1. Using three colors code the length, width, and height of the boxes such that the colors converge at one corner.
2. Label each box *A, B, C,* etc. for identification. The pair of boxes with the same dimensions should be color coded differently (i.e., the longest side could be red on one box and green on the other box).

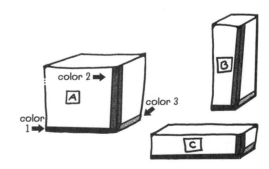

3. Divide the class into groups and then either rotate the students or the boxes so that all eight boxes are observed by all groups.
4. To facilitate calculations for the less experienced students, all measurements could be rounded off to the nearest centimeter. More experienced students could use fractions, millimeters, or decimals (i.e., 3 3/10 cm, 33 mm, 3.3 cm).
5. Students can use colored pencils (the same colors as the construction paper used to identify length, width, and height) to record the various measures on the chart.

## Procedure
1. Place all boxes so that only one colored edge is vertical.
2. While the students are examining the boxes ask, "Which color represents the height?" (Unless you followed the same color scheme on all boxes, the students will be giving different responses. You may need to inform the students that the box rests on its base. The longer base dimension is then called the length, and the shorter side is called the width. The height is the side perpendicular to the base.)
3. Once the dimensions have been defined and identified, have the students complete the chart.

## Connecting Learning
1. Did all lengths, widths, and heights have the same color?
2. Were there any equal dimensions with different colors? How is this explained?
3. Does the order of the numbers matter when multiplying to find the volume? [No.]

## Extension
After the students have completed their work, rotate the boxes changing the position of the colors in relation to the table. Then ask the students to identify the new dimensions.

* Reprinted with permission from *Principles and Standards for School Mathematics,* 2000 by the National Council of Teachers of Mathematics. All rights reserved.

## Key Question

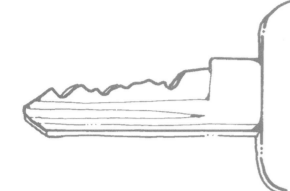

What is the length, width, and height of each box?

## Learning Goal

### Students will:

measure the lengths, widths, and heights of various boxes to determine their volumes.

# WHICH WAY IS UP?

| Box | length | | width | | height | | Volume |
|-----|--------|--------|-------|--------|--------|--------|--------|
| | Color | Measure | Color | Measure | Color | Measure | |
| A | | | | | | | |
| B | | | | | | | |
| C | | | | | | | |
| D | | | | | | | |
| E | | | | | | | |
| F | | | | | | | |
| G | | | | | | | |
| H | | | | | | | |

1. Measure and record the three dimensions of each box.

2. Find the volume of each box.

## Connecting Learning

1. Did all lengths, widths, and heights have the same color?

2. Were there any equal dimensions with different colors? How is this explained?

3. Does the order of the numbers matter when multiplying to find the volume?

11

## Topic
Volume of cylinder

## Key Question
How much water will each container hold?

## Learning Goal
Students will estimate the volume of various cylinders and check their estimates by measuring the amount of water each holds.

## Guiding Documents
*Project 2061 Benchmarks*
- *Use, interpret, and compare numbers in several equivalent forms such as integers, fractions, decimals, and percents.*
- *Use numerical data in describing and comparing objects and events.*

*NCTM Standard 2000\**
- *Work flexibly with fractions, decimals, and percents to solve problems*

## Math
Estimating
　　rounding
Measuring
　　volume
Ordering
Rational numbers
　　percents

## Integrated Processes
Observing
Classifying
Collecting and recording data
Interpreting data
Generalizing
Applying

## Materials
Empty pill bottles and cans of varying heights and
　　diameters
Graduated cylinders
Water
Metric rulers

## Background Information
　　Density is the measure of mass per unit volume. It is defined by the formula $D = m/v$. Students will need to have an intuitive feeling for volume. In cylinders, this is most difficult. This investigation should give them some experiences with cylinders and volume. They will learn that you must consider both the height and diameter of the cylinder which will lead them to the formula of the volume of a cylinder which is $(\pi) r^2 h$ (see *Tin Can Space*).

## Management
1. This investigation can be done with the class as a whole unit or as an independent unit for groups. It can also be set up as a station with students rotating through and sharing their data after everyone has completed the measurements.
2. You will only need 15 to 20 minutes for this investigation.
3. You will need a minimum for seven containers. Try to find some that are tall and slender and others that are short and squat.
4. Assign each cylinder a letter for identification.

## Procedure

1. Ask the students to rank order the cylinders according to the volume they hold by writing down the letters of the cylinders from the smallest volume to the greatest volume.
2. Ask the students to estimate the volume (in mL) of each cylinder and record.
3. Have a student fill each cylinder with water and then pour the contents into a graduated cylinder to find the volume.
4. Put the cylinders in order according to volume. Have students check their predicted order against the actual order.
5. Have the students find the difference between the actual and their estimate.
6. When all groups are finished, have them follow the instructions on their activity page to find the percent of error.

## Connecting Learning

1. Which held more water than you thought? Which held less?
2. Did the tallest one have the greatest volume? Do you think this will always be the case?
3. What dimensions do you have to pay attention to when determining volume? [height of the cylinder, and the area of its base]
4. When you figure the volume of a rectangular solid you multiply length times width times height. Do you have height in a cylinder? [yes] Do you have length and width? [You have area which in the rectangular solid is length times width. In this case it would be area of the circular base.]

## Extensions

1. Challenge the students to find cylinders that would have volumes equal to, greater than, or less than those in this activity.
2. Do *Tin Can Space.*

\* Reprinted with permission from *Principles and Standards for School Mathematics,* 2000 by the National Council of Teachers of Mathematics. All rights reserved.

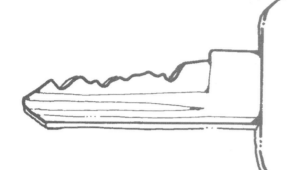

**Key Question**

## How much water will each container hold?

**Learning Goal**

**Students will:**

estimate the volume of various cylinders and check their estimates by measuring the amount of water each holds.

14

| CYLINDER | ESTIMATE OF VOLUME | ACTUAL VOLUME | DIFFERENCE |
|----------|---------------------|---------------|------------|
|          |                     |               |            |
|          |                     |               |            |
|          |                     |               |            |
|          |                     |               |            |
|          |                     |               |            |
|          |                     |               |            |
|          |                     |               |            |

## TO FIND % ERROR:

1. Find the difference between your estimate and the actual volume.
2. Divide this answer by the actual volume and round to the nearest hundredth.
3. Move the decimal point two places to the right and add the percent sign. You're done!

| CYLINDER | % ERROR MY GUESS | CLASS AVERAGE GUESS | % ERROR OF CLASS GUESS |
|----------|-------------------|----------------------|-------------------------|
|          |                   |                      |                         |
|          |                   |                      |                         |
|          |                   |                      |                         |
|          |                   |                      |                         |
|          |                   |                      |                         |
|          |                   |                      |                         |

## Connecting Learning

1. Which held more water than you thought? Which held less?

2. Did the tallest one have the greatest volume? Do you think this will always be the case?

3. What dimensions do you have to pay attention to when determining volume?

4. When you figure the volume of a rectangular solid, you multiply length times width times height. Do you have a height in a cylinder? Do you have length and width?

**Topic**
Volume

**Key Question**
How accurately can you calculate and measure volume?

**Learning Goal**
Students will estimate, calculate, and measure the volume of six different tin cans.

**Guiding Documents**
*Project 2061 Benchmarks*
* *Usually there is no one right way to solve a mathematical problem; different methods have different advantages and disadvantages.*
* *Know that often different explanations can be given for the same evidence, and it is not always possible to tell which one is correct.*

*NCTM Standards 2000\**
* *Select and apply techniques and tools to accurately find length, area, volume, and angle measures to appropriate levels of precision*
* *Work flexibly with fractions, decimals, and percents to solve problems*
* *Develop and use formulas to determine the circumference of circles and the area of triangles, parallelograms, trapezoids, and circles and develop strategies to find the area of more-complex shapes*
* *Develop strategies to determine the surface area and volume of selected prisms, pyramids, and cylinders*

**Math**
Whole number operations
Measuring
    linear
    volume
Using formulae
    area of a circle
    volume of a cylinder
Using rational numbers
    decimals
    percents
Estimating

**Integrated Processes**
Observing
Collecting and recording data
Interpreting data
Applying

**Materials**
Cans of assorted sizes
Metric rulers with mm markings
Graduated cylinders
Water

**Background Information**
For both rectangular solids and cylinders, volume is calculated area x height. With a cylinder, students should be aware that the area measurement is that of its circular base. Because area measurements are given in squared units and then multiplied by height units to obtain the volume, the answer should be reported in cubic units—(unit x unit) x unit = cubic units. Cubic centimeters can be written *cc* or $cm^3$. Students also need to further understand that one cubic centimeter equals one milliliter.

Volume of a cylinder = Area of circle x height of cylinder
    Area of a circle = $\pi$ x $r^2$

**Management**
1. The estimated time for this activity is two class periods.
2. A week prior to this activity, you may want to ask students to bring in some empty tin cans.
3. Each group needs six cans. If you do not have enough cans, they can be rotated through the various groups.

4. Remove all labels and number the cans so that students can refer back to the same can when verifying calculations by direct measurement of volume.

5. Be sure that small graduated cylinders are available so that students can obtain measurements of liquid to at least one milliliter of accuracy.
6. Calculators may be appropriate for the various calculations.

## Procedure

1. Review what students learned about the volume of cylinders from *Can You Tell?* Remind them that they need two figures to determine the volume of a cylinder, the area of the circular base and the height. Tell them that to find the area of the base, they will need to measure the diameter of the interior of the can to the nearest millimeter and divide by two to get the radius.
2. Go over the formula for calculating the area of a circle, $\pi \times r^2$.
3. Have students measure the height of the can to the nearest millimeter and record.
4. Instruct them to calculate the volume of the can (volume=area x height).
5. Have the students check the volume calculations by filling the can with water and measuring the volume with a graduated cylinder.
6. Direct them to calculate the percent error by the following steps:
    a. subtract the measured volume and the calculated volume (smaller from larger)
    b. divide by measured volume
    c. multiply step *b* by 100 and add percent sign.

## Connecting Learning

1. What could account for your calculated errors? [Some error is built-in by limitations of accuracy of the equipment. Some measurements may have been made from the outside of the can so there was an error caused by the thickness of the metal and height of the lip of the can. Other sources of error could be careless measurements, spillage, and not filling the can to the top.]
2. Do you think that your percentage of error is acceptable? Explain. (The concept of acceptable margin of error is an important one when students begin to quantify experimental results. Calculation of experimental error is too complex to be introduced here; however, percent error should allow students to develop an understanding that there will always be some small error but that is not an excuse for sloppy work that leads to larger errors. Reasonably careful work should hold the error under 5%.)
3. How does the formula for the volume of a cylinder compare to the formula for the volume of a rectangular solid? [The volume of the cylinder is the area of its base times its height. The volume of a rectangular solid is also the area of its base (length x width) times its height.]
4. What are you wondering now?

## Extension

In place of checking volume by measuring, fill the can with water and find the mass using an accurate balance. Subtract the mass of the empty can from the mass of the full can. Since 1 cc has an approximate mass of 1 g, this will give a fairly good indication of volume.

\*    Reprinted with permission from *Principles and Standards for School Mathematics,* 2000 by the National Council of Teachers of Mathematics. All rights reserved.

TIN CAN SPACE

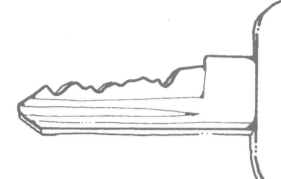

## Key Question

How accurately can you calculate and measure volume?

## Learning Goal

**Students will:**

estimate, calculate, and measure the volume of six different tin cans.

1. Estimate the volume for each of the six different tin cans in cubic centimeters.
2. Find the volume of each by making appropriate measurements and calculations.
3. Check your answers by measuring the volumes using water and a graduated cylinder.
4. Compute your percent of error.

**HERE ARE FORMULAS THAT WILL HELP YOU:**

- Area of Base = $\pi r^2$

- Volume = $\pi r^2 h$

- Percent of Error $= \dfrac{\text{Difference between calculated and measured volume}}{\text{measured volume}}$

| Can Number | Estimated Volume | Diameter (cm) | Radius (cm) | Area of Base (cm²) | Height (cm) | Calculated Volume (cm³) | Measured Volume (cm³) |
|---|---|---|---|---|---|---|---|
| | | | | | | | |
| | | | | | | | |
| | | | | | | | |
| | | | | | | | |
| | | | | | | | |
| | | | | | | | |

## Percent of Error

| Can Number | Difference between Estimated & Measured | Percent of Error | Difference between Calculated & Measured | Percent of Error |
|---|---|---|---|---|
| | | | | |
| | | | | |
| | | | | |
| | | | | |
| | | | | |
| | | | | |

OBSERVATIONS:

## Connecting Learning

1. What could account for your calculated errors?

2. Do you think that your percentage of error is acceptable? Explain.

3. How does the formula for the volume of a cylinder compare to the formula for the volume of a rectangular solid?

4. What are you wondering now?

# A Displaced Object

## Topic
Water displacement

## Key Question
How can you find the volume of irregularly-shaped objects?

## Learning Goal
Students will use the water displacement method for finding the volume of irregularly-shaped objects.

## Guiding Documents
*Project 2061 Benchmark*
- *Usually there is no one right way to solve a mathematical problem; different methods have different advantages and disadvantages.*

*NCTM Standards 2000\**
- *Solve problems that arise in mathematics and in other contexts*
- *Apply and adapt a variety of appropriate strategies to solve problems*

## Math
Measuring
    volume
Whole number operations

## Science
Physical science
    water displacement

## Integrated Processes
Observing
Collecting and recording data
Interpreting data
Generalizing
Applying

## Materials
Variety of classroom objects which are irregular in shape and will not be harmed by submersion into water, such as single hole paper punch, small stapler, set of keys, ink pens, staple remover, glue bottle, etc.
Graduated cylinders large enough to accommodate collected objects

## Background Information
Volume is the amount of space something occupies. When asked how to determine volume, students will most frequently say to multiply length by width by height. While this works for rectangular solids, it is not an appropriate method for finding the volume of irregularly shaped objects. Because an object that is submerged in water will displace a volume of water equal to its volume, the water displacement method is often used to determine the volume of irregularly-shaped objects.

To use the *water displacement method*, start with a determined water level in a graduated cylinder, submerge the object in the water, read the level of the water, and find the difference in the two levels of water. This difference is the volume of the object.

## Management
1. This activity should be done with students in small groups.
2. Because there may be a limited supply of objects, the objects can be rotated through the groups. Groups can compare their results with those of other groups.
3. The objects will need to be submerged in the water in order to achieve an accurate reading of volume. If the objects don't go under the water level on their own, have the students use a pencil to force them under the water level. Caution them to push the object just below the water level so that the pencil does not influence the volume reading.
4. When students add water to the graduated cylinder, they will need to use enough that the complete object can be submerged.

## Procedure

1. Have students estimate and record the volume of the irregularly-shaped objects.
2. Explain the procedure of finding volume using the water displacement method.

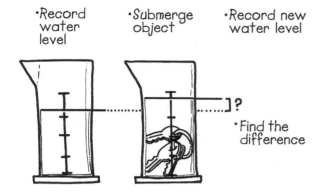

•Record water level   •Submerge object   •Record new water level

]?

•Find the difference

3. Allow students time to determine the volume of at least five objects.
4. Direct them to compare their results with the results of other groups in the class.

## Connecting Learning

1. How did your results compare with the results of other groups? What reasons can you think of for your differences?
2. Do you think this method will work for all objects? Explain.
3. What would happen if you tried this method with a sponge?
4. What applications can you think of in which this method would come in handy?

## Extensions

1. Have students find a rectangular solid and figure the volume using the formula $V = l \times w \times h$. Then have them use the water displacement method and compare the results. (One measure will be in cm$^3$, the other in mL. Make certain that the students realize 1 mL of water (4° C) equals 1 cubic centimeter.)
2. Have the students calculate the percent error between their estimate and the actual volume.

# A Displaced Object

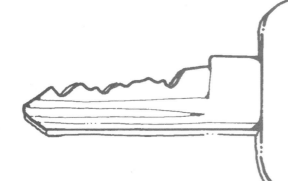

**Learning Goal**

## Students will:

use the water displacement method for finding the volume of irregularly-shaped objects.

# A Displaced Object

1. Estimate the volume of each object.
2. Put water in a graduated cylinder. Record the level of water.
3. Add an object to the water. Record the new water level.
4. Find the difference in the two water levels. This is the volume of your object.
5. Repeat this procedure with all the objects.

| Object | Estimated Volume (mL) | Beginning Water Level (mL) | Ending Water Level (mL) | Difference in Water Levels (mL) | Volume of Objects (mL) |
|--------|----------------------|----------------------------|-------------------------|--------------------------------|------------------------|
|        |                      |                            |                         |                                |                        |
|        |                      |                            |                         |                                |                        |
|        |                      |                            |                         |                                |                        |
|        |                      |                            |                         |                                |                        |
|        |                      |                            |                         |                                |                        |
|        |                      |                            |                         |                                |                        |
|        |                      |                            |                         |                                |                        |

Explain why this method works for finding the volume of these objects.

# A Displaced Object

## Connecting Learning

1. How did your results compare with the results of other groups? What reasons can you think of for your differences?

2. Do you think this method will work for all objects? Explain.

3. What would happen if you tried this method with a sponge?

4. What applications can you think of in which this method would come in handy?

# FILL'ER UP *GRAVEL*

## Topic
Pore space in gravel

## Key Question
How much air space exists between the particles of gravel?

## Learning Goal
Students will determine the pore space in gravel by measuring the volume of water that can be added to it.

## Guiding Documents
*Project 2061 Benchmarks*
- *Use, interpret, and compare numbers in several equivalent forms such as integers, fractions, decimals, and percents.*
- *Find the mean and median of a set of data.*

*NCTM Standards 2000\**
- *Understand and use ratios and proportions to represent quantitative relationships*
- *Work flexibly with fractions, decimals, and percents to solve problems*

## Math
Averaging
Rational numbers
   ratio
   percent
Measuring
   volume
Estimating

## Science
Earth science
   pore space

## Integrated Processes
Observing
Collecting and recording data
Interpreting data
Applying
Generalizing

## Materials
*For a class of 30:*
   four to five liters of dry gravel
   graduated cylinders
   funnels, optional
   water
   15 small empty cans (100 mL minimum)
   pieces of cardboard or tagboard, 5 x 10 cm

## Background Information
It is essential that students know how to use a graduated cylinder with metric units (mL) if results are to be accurate.

The activity sheet takes students through step-by-step procedures for calculating the various ratios and percentages.

In order to express the ratio of water to gravel:

$$\frac{\text{Avg. volume of water (mL)}}{\text{Avg. volume of gravel (mL)}} = \text{Ratio of water to gravel}$$

(The volume of air space displaced is the same as the volume of water used.)

To find the percent of space found in any given volume of gravel:

$$\frac{\text{Avg. volume of water x 100}}{\text{Avg. volume of gravel}} = \text{Percent of space}$$

To compute the average of gravel and of water used:

$$\frac{\text{Trial \#1 + Trial \#2 + Trial \#3}}{3} = \text{Average}$$

## Management
1. Time needed: One to two 45-minute periods.
2. Groups of two work best.
3. To reduce the amount of movement and confusion, have only one of the lab partners get the gravel, water, etc., and perform the measuring tasks while the other student records the data. Since three trials are performed, the partners can switch their roles to gain experience in both.

4. Put the dry gravel in four or five separate containers. Place them on a large table or counter and near a water source. Provide a small scoop for each container. Have your graduated cylinders, tin cans, cardboard, and funnels in the same area.

5. If you are using graduated cylinders, large necked funnels should be used to avoid spilling the gravel.
6. Provide a large container for the wet gravel at the completion of this investigation.

## Procedure

1. Ask students to record their estimate of the volume of air space they think will exist in 100 mL of gravel.
2. Direct them to fill a graduated cylinder with 100 mL of dry gravel. Stress that they be as accurate as possible.

100 mL gravel

100 mL water

add to gravel

3. Have them pour the gravel into a tin can, without packing it. Direct them to use the cardboard to carefully level the gravel on top.
4. Have the students fill the graduated cylinder to the 100 mL mark with water.
5. Ask them to carefully *and slowly* pour the water into the tin can until it fills to the top (but not over) the level of the gravel.
6. Have the students record the volume of water used.
7. Direct the students to repeat steps 2-6 for trials two and three.
8. Have them find and record the average amount of gravel and water used.
9. Have students follow the instructions to determine the ratio of water to gravel as taken from the total and average rows.
10. They should then calculate the volume of air space displaced and the percent of air in gravel.

## Connecting Learning

1. Could you see that space exists between the particles of gravel without having to do this investigation?
2. For what reasons, then, did we do this investigation?
3. How can we apply the methods learned in this investigation to find if space exists in other forms of matter? Name some of forms of matter to which we could apply these methods.
4. How did your results compare with other groups? Should the *percent of air space in gravel* be the same for all groups? Explain.
5. What are you wondering now?

## Extensions

1. Make a list of other items in which you could find the amount of space using the method learned here. Examples would be: different grades of gravel, soils, marbles, rice, beans, etc.
2. Students could devise ways to find the amount of space in other items such as: a sponge, balsa wood, cork, fabric, etc.
3. See the investigation *Fill'er Up Sand.*

* Reprinted with permission from *Principles and Standards for School Mathematics,* 2000 by the National Council of Teachers of Mathematics. All rights reserved.

# FILL'ER UP GRAVEL

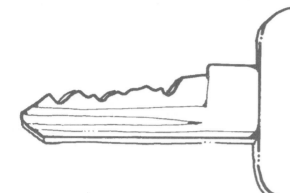

## Key Question

How much air space exists between the particles of gravel?

## Learning Goal

### Students will:

determine the pore space in gravel by measuring the volume of water that can be added to it.

Dry Gravel

Wet Gravel

29

# Fill'er Up GRAVEL

1. Estimate the volume of air space (mL) found in 100 mL of gravel.
2. Measure 100 mL of gravel in a graduated cylinder.
3. Pour gravel into a tin can. Don't pack it! Level the top.
4. Pour 100 mL water into the graduated cylinder.
5. Pour water from the graduated cylinder into the gravel until it just reaches the top of the gravel.
6. Record volume of water used.
7. Repeat steps 2-6 for trials 2 and 3.

Estimate:

100 mL gravel contains _____ mL air space.

| TRIAL NUMBER | VOLUME of GRAVEL (mL) | VOLUME of WATER (mL) |
|---|---|---|
| 1 | | |
| 2 | | |
| 3 | | |
| TOTAL | | |
| AVERAGE | | |

Volume of Air Space Displaced _____
(same as volume of water used)

**Ratio of Water to Gravel**
Formula:

$$\frac{\text{Avg. Volume of Water (mL)}}{\text{Avg. Volume of Gravel (mL)}} = \text{Ratio of Water to Gravel}$$

**Percent of Air Space in Gravel** _____
Formula:

$$\frac{\text{Avg. Volume of Water}}{\text{Avg. Volume of Gravel}} \times 100 = \text{percent of space}$$

Difference Between My Estimated Volume of Space and the Actual (mL)

_____

# FILL'ER UP *GRAVEL*

## Connecting Learning

1. Could you see that space exists between the particles of gravel without having to do this investigation?

2. For what reasons, then, did we do this investigation?

3. How can we apply the methods learned in this investigation to find if space exists in other forms of matter? Name some forms of matter to which we could apply these methods.

4. How did your results compare with other groups? Should the *percent of air space in gravel* be the same for all groups? Explain.

5. What are you wondering now?

## Topic
Space in sand

## Key Questions
1. Does space exist in a substance that appears to not contain any?
2. If space does exist, how do we find out?

## Connecting Learning
Students will determine the percent of space existing in sand.

## Guiding Documents
*Project 2061 Benchmarks*
- *Use, interpret, and compare numbers in several equivalent forms such as integers, fractions, decimals, and percents.*
- *Find the mean and median of a set of data.*
- *Organize information in simple tables and graphs and identify relationships they reveal.*

*NCTM Standards 2000\**
- *Understand and use ratios and proportions to represent quantitative relationships*
- *Work flexibly with fractions, decimals, and percents to solve problems*
- *Collect data using observations, surveys, and experiments*
- *Represent data using tables and graphs such as line plots, bar graphs, and line graphs*

## Math
Averaging
Rational numbers
   ratio
   percent
Graphing
Estimating

## Science
Earth science
   pore space

## Integrated Processes
Observing
Predicting
Collecting and recording data
Interpreting data
Applying
Generalizing

## Materials
*For a class of 30:*
   4-5 liters of dry sand
   graduated cylinders
   funnels, optional
   water
   15 small empty cans, 100 mL minimum
   3" x 5" cards
   completed activity pages from *Fill'er Up Gravel*

## Background Information
   This activity should follow *Fill'er Up Gravel*. While the pore spaces between the gravel particles were quite evident, students may be surprised in this investigation to find that pore spaces exist between the smaller particles of sand. The procedure is very similar to that used in *Fill'er Up Gravel*, but in addition to the calculations, this investigation utilizes a graph to compare the *percent of air space* of the gravel and sand.

   In order to express the ratio of water to sand:

$$\frac{\text{Avg. volume of water (mL)}}{\text{Avg. volume of sand (mL)}} = \text{Ratio of water to sand}$$

(The volume of air space displaced by water is the same as the volume of water used.)

   To find the percent of space existing in any given volume of sand:

$$\frac{\text{Avg. volume of water (mL)}}{\text{Avg. volume of sand (mL) x 100}} = \text{Ratio of water to sand}$$

   To compute the average volume of sand and of water used:

$$\frac{\text{Trial \#1 + Trial \#2 + Trial \#3}}{3} = \text{Average}$$

It is important to convey to the students the idea that we are investigating to see if we can answer the *Key Question*. The basic differences between *Fill'er Up Gravel* and *Fill'er Up Sand* is that spaces could be seen between the particles of gravel but cannot be seen in sand. This activity and *Fill'er Up Gravel* utilize a process called the *water displacement method* in which water displaces the air spaces in the sand and gravel. The quantity (volume) of water used is therefore the quantity (volume) of air space which exists in the sand and gravel.

## Management
1. Time needed: One to two 45-minute periods.
2. Groups of two work best.
3. To reduce the amount of movement and confusion, have only one of the lab partners get the sand, water, etc., and perform the measuring tasks while the other student records the data. Since three trials are performed, the partners can switch their roles to gain experience in both.
4. Put the dry sand in four or five separate containers. Place them on a large table or counter and near a water source. Provide a small scoop for each container. Have your graduated cylinders, tin cans, cardboard, and funnels in the same area.
5. If you are using graduated cylinders, large necked funnels should be used to avoid spilling the sand.
6. Provide a large container for the wet sand at the completion of this investigation.

## Procedure
1. Ask students to record their estimates of the volume of air space they think will exist in 100 mL of sand.
2. Direct them to fill a graduated cylinder with 100 mL of dry sand. Stress that they be as accurate as possible.
3. Have them pour the sand into a tin can, without packing it. Direct them to use the cardboard to carefully level the sand on top.
4. Have the students fill the graduated cylinder to the 100 mL mark with water.
5. Ask them to *slowly* pour the water into the tin can until it fills to the top (but not over) the level of the sand.
6. Have the students record the volume of water used.
7. Direct the students to repeat steps 2-6 for trials two and three.
8. Have them find and record the average amount of sand and water used.
9. Have students follow the instructions to determine the ratio of water to sand as taken from the total and average rows.

10. They should then calculate the volume of air space displaced and the percent of air in sand.
11. Have students complete the graphs to compare the *Volume of Air Displaced* and the *Percent of Air Space* for gravel (*Fill'er Up Gravel*) and sand.

## Connecting Learning
1. Sand did not appear to contain any space between its particles and yet we proved that space does exist there. Do you think this shows that substances contain space even though this space might be invisible? Why or why not?
2. How did the pore space in gravel compare to the pore space in sand?
3. What do the two graphs show you?
4. What conclusions can you draw from these two activities?
5. What are you wondering now?

## Extensions
1. List substances for which we *could not* use the water displacement method to see if they contain space. Examples: solid metals, plastics, any non-porous materials.
2. List substances that might contain "invisible" space to which we could apply the water displacement method.
3. Devise a method to find out if space exists in a liquid like water or alcohol.

\* Reprinted with permission from *Principles and Standards for School Mathematics,* 2000 by the National Council of Teachers of Mathematics. All rights reserved.

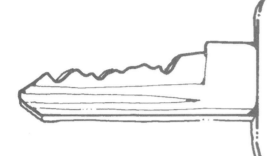 *Fill'er Up*  SAND

## Learning Goal

**Students will:**

determine the percent of space existing in sand.

# Fill'er Up SAND

1. Estimate the volume of air (mL), if any, found in 100 mL of sand.
   **100 mL sand contains _____ mL air space.**
2. See Fill'er Up Gravel instructions #2-7, but use sand instead.
3. Graph to compare the volume of air space displaced in gravel and sand.
4. Graph to compare the percent of air space in gravel and sand.

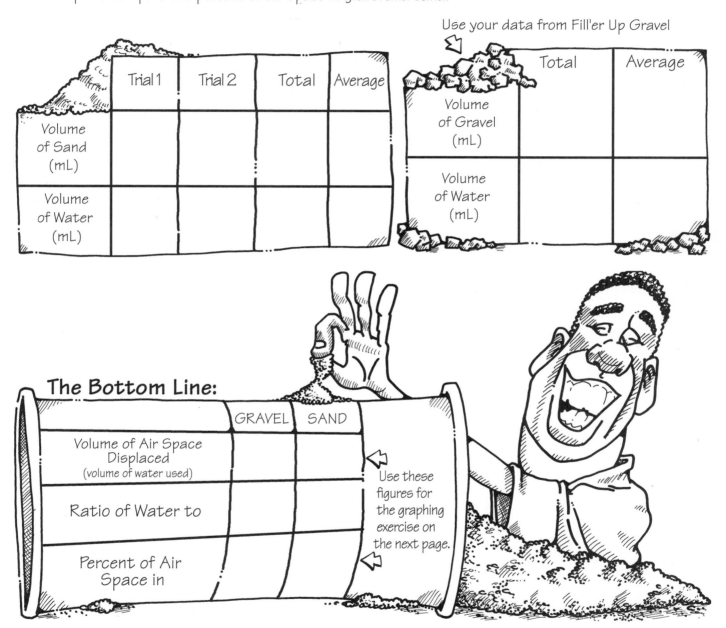

Use your data from Fill'er Up Gravel

|  | Trial 1 | Trial 2 | Total | Average |
|---|---|---|---|---|
| Volume of Sand (mL) |  |  |  |  |
| Volume of Water (mL) |  |  |  |  |

|  | Total | Average |
|---|---|---|
| Volume of Gravel (mL) |  |  |
| Volume of Water (mL) |  |  |

## The Bottom Line:

|  | GRAVEL | SAND |
|---|---|---|
| Volume of Air Space Displaced (volume of water used) |  |  |
| Ratio of Water to |  |  |
| Percent of Air Space in |  |  |

Use these figures for the graphing exercise on the next page.

Difference between my estimated volume of space in sand and the actual _____ mL

On the back of this paper, write what you've learned about space in matter.

# Fill'er Up

SAND

## Graph to compare

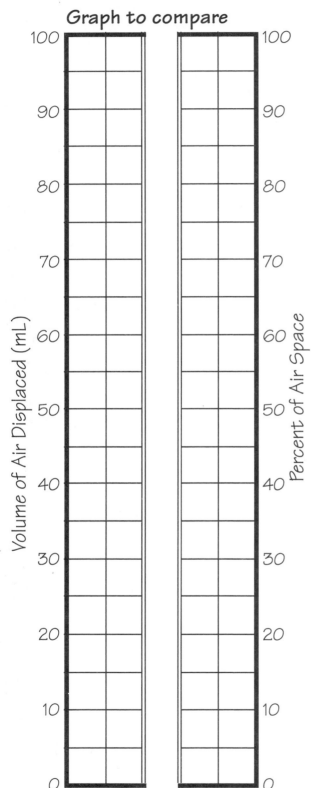

Volume of Air Displaced (mL)

100
90
80
70
60
50
40
30
20
10
0

Gravel  Sand

Percent of Air Space

100
90
80
70
60
50
40
30
20
10
0

Gravel  Sand

Write two conclusions you can make by looking at the graphs.

# *Fill'er Up* SAND

## Connecting Learning

1. Sand did not appear to contain any space between its particles and yet we proved that space does exist there. Do you think this shows that substances contain space even though this space might be invisible? Why or why not?

2. How did the pore space in gravel compare to the pore space in sand?

3. What do the two graphs show you?

4. What conclusions can you draw from these two activities?

5. What are you wondering now?

# A Salty Solution

## Topic
Space in water

## Key Question
What will happen to a volume of water if salt is added to it?

## Learning Goal
Students will investigate whether space exists even among the molecules that make up water.

## Guiding Documents
*Project 2061 Benchmarks*
- *Use, interpret, and compare numbers in several equivalent forms such as integers, fractions, decimals, and percents.*
- *What people expect to observe often affects what they actually do observe. Strong beliefs about what should happen in particular circumstances can prevent them from detecting other results. Scientists know about this danger to objectivity and take steps to try and avoid it when designing investigations and examining data. One safeguard is to have different investigators conduct independent studies of the same questions.*

*NCTM Standards 2000\**
- *Understand and use ratios and proportions to represent quantitative relationships*
- *Work flexibly with fractions, decimals, and percents to solve problems*
- *Collect data using observations, surveys, and experiments*

## Math
Averaging
Rational numbers
    ratios
    percent
Measuring
    volume
Estimating

## Science
Observing
Predicting
Collecting and recording data
Applying
Generalizing

## Materials
*Per group:*
    clear drinking glass
    50 mL iodized salt
    one graduated cylinder
    water
    one non-porous stirring stick (no wood)
    fine-tipped felt pen

## Background Information
Space exists between the molecules that make up water. When salt is dissolved in water, it takes up some of that space. Students will find that a small amount of salt can be added to water without changing its level. Volumes are not additive. The salt fills in some of the spaces in the water.

## Management
1. The suggested time is one to two 45-minute periods.
2. Students should work in pairs.
3. One partner should gather materials and perform the hands-on tasks while the other records the data. Partners can switch responsibilities for added experience.
4. All materials should be laid out near the water source.
5. Felt pen markings on the drinking glass should be as fine as possible.

6. Water should be at room temperature.
7. Students should be made aware that, when adding salt, add it *very slowly and carefully*. If too much salt is added at one time, results may not be accurate.
8. Make sure that stirring tools are absolutely non-porous. A porous material will soak up water affecting the accuracy of the outcome.

## Procedure

1. Instruct students to pour 100 mL of water into a glass and mark the water level on the glass with a fine-tipped marking pen.
2. Using a dry graduated cylinder, have them measure 50 mL of salt.

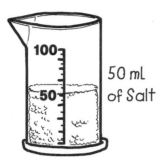

50 mL of Salt

3. Direct the students to carefully pour 1 mL of salt into the water, stir and dissolve. Have them check the water level. Instruct them to continue this process until they observe the water level begin to rise at which time they should stop adding salt.

Slowly add 1 mL of salt

4. Have the students record the amount of salt used and repeat steps 1-4 for two more trials.
5. Inform the students that they will continue the same procedure but use 250 mL of water.
6. Direct them to calculate the average amount of salt used with 100 mL of water and again with 250 mL of water.
7. They should then calculate the ratio of salt to water for both solutions and the percentage of salt in each final solution.

## Connecting Learning

1. Why were we able to add some salt to the water without raising its level? [The salt filled the spaces in water.]
2. What other materials do you think would produce similar results?
3. What materials could not be used? Why?
4. What basic change in the added material must take place for the results of this investigation to hold true? [The material must dissolve.]

## Extensions

1. Try doing this investigation using non-iodized salt, rock salt. Compare the results with those of iodized salt.
2. Do this investigation using different liquids like milk, juices, alcohol, etc. Try these new liquids with the different kinds of salts. Compare results.
3. Try this investigation using various temperatures of water. Try to determine if the amount of space changes with the change in temperature.
4. For further study, see *A Super Salty Solution* found in this book.

\* Reprinted with permission from *Principles and Standards for School Mathematics,* 2000 by the National Council of Teachers of Mathematics. All rights reserved.

# A Salty Solution

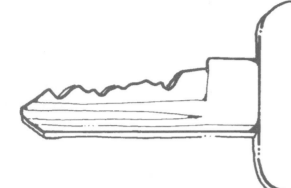

**Key Question**

What will happen to a volume of water if salt is added to it?

**Learning Goal**

**Students will:**

investigate whether space exists even among the molecules that make up water.

# A Salty Solution

1. Pour 100 mL of water into a glass. Mark the water level on the glass with a marking pen.
2. Using a dry graduated cylinder, measure 50 mL of salt.
3. Carefully pour one mL of salt into the water, stir until dissolved. Check the water level. Continue this process until you see the water level begin to rise. Stop adding salt when the water level begins to rise.
4. Record the amount of salt used.
5. Repeat steps 1-4 for trials 2 and 3
6. Repeat steps 1-5 using 250 mL of water.
7. Calculate the average amount of salt used with 100 mL of water and again with 250 mL of water.
8. Calculate the ratio of salt to water for both solutions.

$$\text{Ratio} = \frac{\text{average mL of salt}}{\text{mL of water}}$$

9. Calculate the percentage of salt in each final solution.

$$\text{Percentage} = \frac{\text{average mL of salt}}{\text{mL of water}} \times 100$$

| | 100 mL of water | | 250 mL of water |
|---|---|---|---|
| **SALT USED** | Trial 1 | | Trial 1 |
| | Trial 2 | | Trial 2 |
| | Trial 3 | | Trial 3 |
| | Total | | Total |
| Average | | Average | |
| Ratio | | Ratio | |

# A Salty Solution

## Connecting Learning

1. Why were we able to add some salt to the water without raising its level?

2. What other materials do you think would produce similar results?

3. What materials could not be used? Why?

4. What basic change in the added material must take place for the results of this investigation to hold true?

# Wat-ar DENSITIES?

## Topic
Volume and density

## Key Question
What is the density of water?

## Learning Goal
Students will determine the density of water.

## Guiding Documents
*Project 2061 Benchmarks*
- *Organize information in simple tables and graphs and identify relationships they reveal.*
- *Find the mean and median of a set of data.*
- *Decide what degree of precision is adequate and round off the result of calculator operations to enough significant figures to reasonably reflect those of the inputs.*
- *Graphs can show a variety of possible relationships between two variables. As one variable increases uniformly, the other may do one of the following: always keep the same proportion to the first, increase or decrease steadily, increase or decrease faster and faster, get closer and closer to some limiting value, reach some intermediate maximum or minimum, alternately increase and decrease indefinitely, increase and decrease in steps, or do something different from any of these.*

*NCTM Standards 2000\**
- *Use observations about differences between two or more samples to make conjectures about the populations from which the samples were taken*
- *Select, create, and use appropriate graphical representations of data, including histograms, box plots, and scatterplots*
- *Solve simple problems involving rates and derived measurements for such attributes as velocity and density*
- *Collect data using observations, surveys, and experiments*

## Math
Measuring
   volume
   mass
Whole number operations
Graphing
Estimation
   rounding

## Science
Physical science
   density

## Integrated Processes
Observing
Collecting and recording data
Interpreting data
Applying
Generalizing

## Materials
Graduated cylinders
300 mL of water (room temperature) per group
Balances
Masses

## Background Information
Density is defined as mass per unit volume. It is expressed in the units of grams/milliliters or grams/cubic centimeters. The standard density of water is 1 g/cm³. At 4° Celsius, 1 mL of water is the equivalent of 1 cubic centimeter.

## Management
1. Groups should be from two to five students.
2. Allow about two class periods to complete this investigation, depending upon equipment available.

## Procedure

1. Have the students find the mass of the empty graduated cylinder and record this data.
2. Direct them to fill the graduated cylinder to the specified volume and find its mass.
3. If necessary, remind the students that they will have to subtract the mass of the empty graduated cylinder to find the mass of the water.
4. Direct the students to find the density of the measured volume of water by division and rounding off to the nearest hundredth.
5. Have the students graph the mass against the volume of water used. By connecting these ordered pairs, they should produce a straight line.

## Connecting Learning

1. How is density figured? [mass/volume]
2. What are the units used for density in this investigation? [grams and milliliters]
3. Can you write an equation for this graph? ($y = x$, or mass = volume) What is the slope of this line? [1] Can you predict what the mass of 300 mL of water would be? [300 grams]
4. The standard density of water is 1 g/cm³. Was yours the same? If not, why do you think it wasn't? [Our water wasn't at 4° Celsius. We weren't careful enough when we measured the volume of water in the graduated cylinder or our balance and masses weren't accurate enough to distinguish small differences.]
5. What are you wondering now?

## Extensions

1. Try this activity again using ice water or hot water. Do the densities change? [yes]
2. Have your students find the percent error between their density of water and the standard density of 1 g/cm³. To find percent error:
   a. Subtract your density and the standard density
   b. Divide by the standard density
   c. Move your decimal point two places to the right (or multiply by 100) and add a percent sign.

   Example: my density = 1.03 g/cm³
   a. 1.03-1 = .03
   b. .03/1 = .03
   c. .03 x 100 = 3%

* Reprinted with permission from *Principles and Standards for School Mathematics*, 2000 by the National Council of Teachers of Mathematics. All rights reserved.

# Wat-ar DENSITIES?

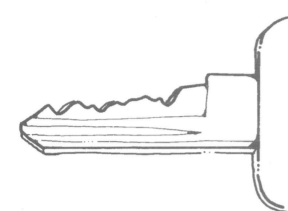

**Key Question**

What is the density of water?

## Learning Goal

**Students will:**

determine the density of water.

# Wat-ar DENSITIES?

1. Record the mass of your empty graduated cylinder on the chart below.
2. Measure 50 mL of water into your graduated cylinder.
3. Record the mass of the graduated cylinder with the water in it.
4. Subtract the two masses to find the mass of the water. Record this under the heading Mass of Water.
5. Divide the mass of the water by the volume of the water to find its density and round to the nearest hundredth.
6. Repeat using the indicated volumes of water.
7. Find the average of the densities.

| Volume | Mass Empty | Mass Filled | Mass of Water | Density of Water |
|--------|-----------|-------------|---------------|------------------|
| 50 mL | | | | |
| 100 mL | | | | |
| 150 mL | | | | |
| 200 mL | | | | |
| 250 mL | | | | |

The Standard Density of Water at room temperature is 1g/cm³ (or 1g/mL).

Average Density =

How close did you come?

# Wat-ar DENSITIES?

Graph your data.

Describe the graphed line.

What does the line tell you about the volume and mass of water?

Mass of water (g)

300

250

200

150

100

50

50   100   150   200   250   300

Volume of water (mL)

# Wat-ar DENSITIES?

## Connecting Learning

1.  How is density figured?

2.  What are the units used for density in this investigation?

3.  Can you write an equation for this graph? What is the slope of this line? Can you predict what the mass of 300mL of water would be?

4.  The standard density of water is 1 $g/cm^3$. Was yours the same? If not, why do you think it wasn't?

5.  What are you wondering now?

# THE ORANGE'S LIFE JACKET

## Topic
Density

## Key Questions
1. How can you find the density of an orange?
2. What can you find out about the densities of the pulp and peel of the orange?

## Learning Goal
Students will find the mass and volume by water displacement to determine the density of oranges.

## Guiding Documents
*Project 2061 Benchmarks*
- *Organize information in simple tables and graphs and identify relationships they reveal.*
- *Offer reasons for their findings and consider reasons suggested by others.*

*NCTM Standards 2000\**
- *Understand and use ratios and proportions to represent quantitative relationships*
- *Collect data using observations, surveys, and experiments*

## Math
Measuring
    volume
    mass
Whole number operations
Using formulae
    density

## Science
Physical science
    density

## Integrated Processes
Observing
Predicting
Controlling variables
Collecting and recording data
Interpreting data
Generalizing
Applying

## Materials
Oranges
Balances
Masses
Large graduated cylinders (large enough so the orange fits inside) or graduated liter box

## Background Information
Density is the measure of mass per unit volume. It is defined by the formula $D = m/v$. When *m (mass)* is measured in grams and *v (volume)* in cubic centimeters, density is expressed in grams per cubic centimeter or milliliters. It can also be thought of as the ratio of mass to volume.

## Management
1. Have students work in groups of three or four.
2. If only one set of materials is available, groups may be rotated through the investigation and results compared when all have finished.
3. The time required for the investigation is about 45 minutes.

## Procedure
1. Set up each station with the materials listed.
2. Have the students use the balances to find the mass of each orange and record the results.
3. Instruct them to fill the graduated cylinder or box with water so the depth is a little less than the diameter of the orange and record this reading.
4. Direct them to place the orange in the water. If it floats, tell them to force it down with the eraser end of a pencil until it is just submerged and read the water depth and record.
5. To determine the volume of the orange, instruct the students to find the difference in the beginning volume and ending volume and record. (If part of the pencil is submerged, a slight error is introduced, so caution the students to just submerge the orange.)
6. Have the students find the density of the orange using the formula $D = m/v$.
7. Have them repeat these steps for four more oranges or gather data from four other groups and find the average density.
8. Ask students what they think will happen to the orange's density if it is peeled. Have them peel the oranges and repeat the above steps to find the average density of peeled oranges.

## Connecting Learning

1. If an orange is placed in water, does it sink or float? Do you think this applies to all oranges?
2. If the orange is peeled, does it sink or float? Does this apply in every case? Explain.
3. Which has the greater density, the orange's pulp or the orange's peel? Explain your answer.
4. What part of the unpeeled orange is above the water line? How did you know this? [From direct observation or from the data table that gives the average density which tells the part that is under the water line. By subtracting the average density from the density of tap water (1.00), you can determine the part that is above the water line.]
5. Is the density of oranges predictable? Explain.
6. Do you think all fruits with peelings will behave the same? Explain.
7. What are you wondering now?

## Extension

Students may wish to investigate the density of apples, bananas, nuts, and other fruits.

* Reprinted with permission from *Principles and Standards for School Mathematics,* 2000 by the National Council of Teachers of Mathematics. All rights reserved.

# THE ORANGE'S LIFE JACKET

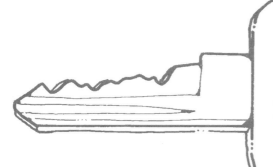

## Key Questions

1. How can you find the density of an orange?
2. What can you find out about the densities of the pulp and peel of the orange?

## Learning Goal

### Students will:

find the mass and volume by water displacement to determine the density of oranges.

# THE ORANGE'S LIFE JACKET

| Unpeeled ORANGE | Mass in grams | < > = | Volume in milliliters | Sink or Float | Ratio: Mass / Volume | Decimal Equivalent |
|---|---|---|---|---|---|---|
| A | | | | | | |
| B | | | | | | |
| C | | | | | | |
| D | | | | | | |
| E | | | | | | |
| Averages | | | | | | |

| Peeled ORANGE | Mass in grams | < > = | Volume in milliliters | Sink or Float | Ratio: Mass / Volume | Decimal Equivalent |
|---|---|---|---|---|---|---|
| A | | | | | | |
| B | | | | | | |
| C | | | | | | |
| D | | | | | | |
| E | | | | | | |
| Averages | | | | | | |

# THE ORANGE'S LIFE JACKET

## Connecting Learning

1. If an orange is placed in water, does it sink or float? Do you think this applies to all oranges? Explain.

2. If the orange is peeled, does it sink or float? Does this apply in every case? Explain.

3. Which has the greater density, the orange's pulp or the orange's peel? Explain your answer.

4. What part of the unpeeled orange is above the water line? How did you know this?

5. Is the density of oranges predictable? Explain.

6. Do you think all fruits with peelings will behave the same? Explain.

7. What are you wondering now?

# THE ORANGE'S SECRET

**Topic**
Density

**Key Questions**
1. How can you find the density of an orange?
2. What can you find out about the densities of the pulp and peel of the orange?

**Learning Goal**
Students will find the mass and volume by water displacement to determine the density of oranges.

**Guiding Documents**
*Project 2061 Benchmarks*
- *Organize information in simple tables and graphs and identify relationships they reveal.*
- *Offer reasons for their findings and consider reasons suggested by others.*

*NCTM Standards 2000\**
- *Understand and use ratios and proportions to represent quantitative relationships*
- *Collect data using observations, surveys, and experiments*
- *Solve simple problems involving rates and derived measurements for such attributes as velocity and density*

**Math**
Measuring
    volume
    mass
Whole number operations
Using formulae
    density

**Science**
Physical science
    density

**Integrated Processes**
Observing
Predicting
Controlling variables
Collecting and recording data
Interpreting data
Generalizing
Applying

**Materials**
Oranges
Balances
Masses
Large graduated cylinders (large enough so the orange fits inside) or graduated liter box

**Background Information**
    Density is the measure of mass per unit volume. It is defined by the formula $D = m/v$. When $m$ *(mass)* is measured in grams and $v$ *(volume)* in cubic centimeters, density is expressed in grams per cubic centimeter or milliliters. It can also be thought of as the ratio of mass to volume.

**Management**
1. Have students work in groups of three or four.
2. If only one set of materials is available, groups may be rotated through the investigation and results compared when all have finished.
3. The time required for the investigation is about 45 minutes.

**Procedure**
1. Set up each station with the materials listed.
2. Have the students use the balances to find the mass of each orange and record the results.
3. Instruct them to fill the graduated cylinder or cube with water so the depth is a little less than the diameter of the orange and record this reading.

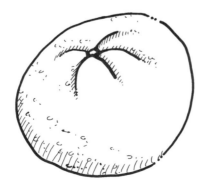

4. Direct them to place the orange in the water. If it floats, tell them to force it down with the eraser end of a pencil until it is just submerged and read the water depth and record.

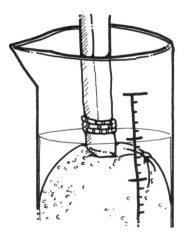

5. To determine the volume of the orange, instruct the students to find the difference in the beginning volume and ending volume and record. (If part of the pencil is submerged, a slight error is introduced, so caution the students to just submerge the orange.)
6. Have the students find the density of the orange using the formula D = m/v.
7. Have them repeat these steps for four more oranges or gather data from four other groups and find the average density.
8. Ask students what they think will happen to the orange's density if it is peeled. Have them peel the oranges and repeat the above steps to find the average density of peeled oranges.

**Connecting Learning**
1. If an orange is placed in water, does it sink or float? Do you think this applies to all oranges?
2. If the orange is peeled, does it sink or float? Does this apply in every case? Explain.
3. Which has the greater density, the orange's pulp or the orange's peel? Explain your answer.
4. What part of the unpeeled orange is above the water line? How did you know this? [From direct observation or from the data table that gives the average density which tells the part that is under the water line. By subtracting the average density from the density of tap water (1.00), you can determine the part that is above the water line.]
5. Is the density of oranges predictable?
6. Do you think all fruits with peelings will behave the same? Explain.
7. What are you wondering now?

**Extension**
Students may wish to investigate the density of apples, bananas, nuts, and other fruits.

\* Reprinted with permission from *Principles and Standards for School Mathematics,* 2000 by the National Council of Teachers of Mathematics. All rights reserved.

# The Orange's Secret

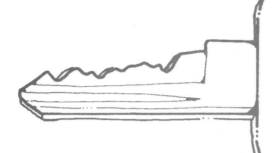

## Key Questions

1. How can you find the density of an orange?
2. What can you find out about the densities of the pulp and peel of the orange?

## Learning Goal

**Students will:**

find the mass and volume by water displacement to determine the density of oranges.

# The Orange's Secret

**Will the oranges float or sink?**
Circle your prediction.

Investigation Team

_____ float   sink
_____ float   sink
_____ float   sink
_____ float   sink
_____ float   sink
_____ float   sink

| Unpeeled Oranges' Data: | A | B | Orange Designations C | D | E |
|---|---|---|---|---|---|
| Mass in grams | | | | | |
| Water depth with orange (mL) | | | | | |
| Water depth without orange (mL) | | | | | |
| Difference in water depth (mL) | | | | | |
| Density | | | | | |
| Average Density | | | | | |

| Peeled Oranges' Data | A | B | C | D | E |
|---|---|---|---|---|---|
| Mass in grams | | | | | |
| Water depth with orange (mL) | | | | | |
| Water depth without orange (mL) | | | | | |
| Difference in water depth (mL) | | | | | |
| Density | | | | | |
| Average Density | | | | | |

How close were your predictions?

What is the average density of unpeeled oranges?

Is it predictable? Explain.

# The Orange's Secret

## Connecting Learning

1. If an orange is placed in water, does it sink or float? Do you think this applies to all oranges?

2. If the orange is peeled, does it sink or float? Does this apply in every case? Explain.

3. Which has the greater density, the orange's pulp or the orange's peel? Explain your answer.

4. What part of the unpeeled orange is above the water line? How did you know this?

5. Is the density of oranges predictable? Explain.

6. Do you think all fruits with peelings will behave the same? Explain.

7. What are you wondering now?

**Topic**
Density

**Key Question**
How can you tell whether an object will sink or float before you put it in water?

**Learning Goal**
Students will determine if objects will float or sink by comparing the mass to volume ratio.

**Guiding Documents**
*Project 2061 Benchmarks*
- *Organize information in simple tables and graphs and identify relationships they reveal.*
- *Use calculators to compare amounts proportionally.*
- *Use, interpret, and compare numbers in several equivalent forms such as integers, fractions, decimals, and percents.*

*NCTM Standards 2000\**
- *Understand and use ratios and proportions to represent quantitative relationships*
- *Collect data using observations, surveys, and experiments*
- *Work flexibly with fractions, decimals, and percents to solve problems*

**Math**
Measuring
    volume
    mass
Ratio and proportion
Rational numbers
    decimals
Using formulae
    density
Equalities and inequalities
    greater than
    less than
    equal to

**Science**
Physical science
    density

**Integrated Processes**
Predicting
Comparing
Recording data
Generalizing
Interpreting

**Materials**
*For each group:*
    graduated cylinder
    10 assorted objects small enough to fit into the
        graduated cylinders (e.g., pencils, ball-point
        pens with cap, glass marbles, iron nuts or bolts,
        rubber erasers, rocks, walnuts or pecans, Teddy
        Bear Counters, etc.)
    balance
    masses
    calculators

**Background Information**
To many, the concept of density is difficult to understand. Some, therefore, skip it in the teaching of science. It need not be this way, especially with a concept that has such far reaching applications in the real world and provides such a wonderful tie between mathematical and scientific concepts. Typically, density is expressed mathematically in the algebraic formula $D = m/v$ where D is the *density*, m is the *mass* and v is the *volume* of the object. However, memorizing a formula is not what the study of density should be about.

Let's look at the formula to see if it can be made easier to understand. Using the term "ratio" is helpful. A ratio is simply a comparison between two numbers. The numbers can represent real-world experiences such as the mass of an object or the amount of space an object occupies (its volume). Density is a comparison between these two observable characteristics of an object. It is an expression of how an object's mass compares to its volume. This becomes significant when we consider whether an object will float or sink in water. The ratio m/v provides the measure of density. Water at 4° Celsius has a density of 1 gram per cubic centimeter. Even at other temperatures, it is so close to 1 gram per cubic centimeter that we can use that value generally. An object with a density greater than 1 gram per cubic centimeter will sink in water and one with a density less than 1 will float (unless its shape or surface tension prevent this).

Take, for example, a swimmer who wishes to float. The swimmer cannot quickly change his mass but can quickly change his volume by inhaling or exhaling. Inhaling causes the volume of the body to increase, increasing v in the denominator of the ratio. Increasing the denominator decreases the value of the ratio and, therefore, the density. If the density is decreased to less than 1 gram per cubic centimeter, the swimmer will float. Conversely, if the swimmer exhales, the volume of the body is decreased, the value of v is decreased causing the value of the ratio m/v to increase. Since the density is increased, the likelihood of sinking increases. Placing the density formula on the board and expanding on this example while observing the formula will help students to relate a real-life activity to its effect in a formula.

There are many misconceptions about how and why things float. Many of us still think that the heavier an object is, the more likely it is to sink. How then do steel ships float? How does a submarine both float and sink? The heaviness of an object, considered by itself, does not play much of a role in floating and sinking. It is the density and buoyancy that are the main factors while surface tension can affect the result in some instances.

The general rule your students will discover is: if an object's mass (in grams) is greater than its volume (in cubic centimeters), it will sink in fresh water. If an object's mass is less than its volume, it will float. Once, students understand the idea of comparing mass to volume, density as a formula becomes less formidable.

Let's take for example an object whose mass is 1 gram and whose volume is 2 cubic centimeters. If we compare these two numbers, we can see that the mass is less than the volume so the object will float. Expressed as a ratio, it is the fraction $1 \text{ g}/2 \text{ cm}^3$. This can be expressed as $.5 \text{g/cm}^3$ or .5 grams per cubic centimeter. (A milliliter is the equivalent of a cubic centimeter.)

Since the density of freshwater is $1 \text{g/cm}^3$, your students will see a pattern starting to develop as they experiment: all items with a decimal equivalent less than 1 float, and all items with a decimal equivalent greater than 1 sink.

To find the volume of an irregularly shaped object, use the *water displacement method*. Simply put the object into a graduated cylinder with a measured volume of water in it, making sure that the entire object is submerged in the water. As the object is submerged, it will be noted that the level of the water rises. This is because the object is now occupying the space where the water was before. The increase in volume, as measured by the rise in the water level, is equal to the volume of the object. For example, if we had a graduated cylinder with 50 mL of water in it and the object made the water level rise to 75 mL, then we would say that the object has a volume of 25 mL or cubic centimeters since that is how much water was displaced. Again, it is important to make sure that the entire object is submerged in the water.

## Management

1. The estimated time for this investigation is one 45-minute class period.
2. Divide the class into groups of 2-4 students.
3. Provide the ten items for the students. It is best to stay away from items such as paper clips, pins, and tacks since their volumes are small and will be hard to measure using the displacement method.

## Procedure

1. Instruct each group to measure and record the mass (in grams) and the volume (in mL) of each object.
2. Have students compare the mass of each object to its volume and place the correct relational symbol in the column provided.
3. Ask students to generate a rule describing the relationship of mass to volume in terms of floating/sinking.
4. Have students record the ratio.
5. Ask students to determine the decimal equivalent.
6. Allow time for students to test whether the objects will actually sink or float.

## Connecting Learning

1. When you compare all the data for the objects that will float, what do you notice? [Their decimal equivalents are less than 1.]
2. What about the data for objects that will sink? [The decimal equivalent is greater than 1.]
3. Why does *one* seem to be the magic number? [It is the density of water. The objects' densities are compared to this.]
4. Were any of your calculations incorrect? How could you tell? [If the decimal equivalent indicated the object would sink (or float) and it did the opposite.] What might be some reasons for error? [Mistakes in reading the instruments; the instruments might not be accurate enough; mistakes in performing the calculations; not submerging the items in the water, etc.]
5. What are you wondering now?

## Extensions

Have students find objects that they place in water. From how these objects behave in the water, have students predict their densities. They should then check them by finding the mass and volume of each object.

* Reprinted with permission from *Principles and Standards for School Mathematics,* 2000 by the National Council of Teachers of Mathematics. All rights reserved.

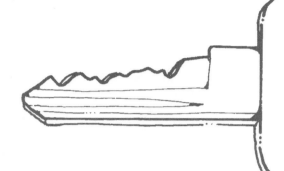

## Key Question

How can you tell whether an object will sink or float before you put it in water?

## Learning Goal

### Students will:

determine if objects will float or sink by comparing the mass to volume ratio.

**How can you tell before you put something in the water, whether it will sink or float?**

| Object | Mass in grams | < > = | Volume in milliliters | Sink or Float | Ratio: Mass / Volume | Decimal Equivalent |
|---|---|---|---|---|---|---|
|  |  |  |  |  |  |  |
|  |  |  |  |  |  |  |
|  |  |  |  |  |  |  |
|  |  |  |  |  |  |  |
|  |  |  |  |  |  |  |
|  |  |  |  |  |  |  |
|  |  |  |  |  |  |  |
|  |  |  |  |  |  |  |
|  |  |  |  |  |  |  |
|  |  |  |  |  |  |  |

How does the relationship of mass to volume let you know if something will sink or float?

## Connecting Learning

1. When you compare all the data for the objects that will float, what do you notice?

2. What about the data for objects that will sink?

3. Why does *one* seem to be the magic number?

4. Were any of your calculations incorrect? How could you tell? What might be some reasons for error?

5. What are you wondering now?

## Topic
Density

## Key Question
What will happen if we drop popcorn kernels, raisins, or birdseed into plain or carbonated water?

## Learning Goal
Students will observe whether or not there is a change in density to objects in regular and carbonated water.

## Guiding Documents
*Project 2061 Benchmarks*
- *Keep records of their investigations and observations and not change the records later.*
- *Keep a notebook that describes observations made, carefully, distinguishes actual observations from ideas and speculations about what was observed, and is understandable weeks or months later.*
- *Make sketches to aid in explaining procedures or ideas.*

*NCTM Standard 2000\**
- *Collect data using observations, surveys, and experiments*

## Integrated Processes
Observing
Making and testing hypotheses
Recording data
Interpreting data
Generalizing
Applying

## Science
Physical science
    density

## Materials
*For each group:*
    tall, narrow jar (the type olives come in)
    water
    seltzer tablets
    small quantities of unpopped popcorn kernels,
        raisins, and/or mixed bird seed

## Background Information
Density is a property of solids, liquids, and gases. We think of it as how heavy a substance is for its size or quantity; for example, we could say that a brick is denser than the same size box full of feathers. Another definition is how much of something is squeezed into a certain-sized space; the more technical scientific definition of density is the mass per unit volume. The temptation is to say heavier rather than denser, but denser is the correct term.

Carbon dioxide ($CO_2$) is a colorless, tasteless, odorless gas. It occurs naturally in the world as a product of our breathing. It is absorbed by plants as part of the photosynthesis process. It is denser than oxygen, but less dense than liquids. In seltzer water and carbonated beverages, the bubbles are $CO_2$. We can produce carbon dioxide in plain water by adding seltzer tablets (sodium bicarbonate).

When a popcorn kernel is dropped into a jar of plain water, it sinks to the bottom of the jar because the corn is denser than the water. A few air bubbles will cling to the corn. When a seltzer tablet is added to the water, bubbles of $CO_2$ are produced. They cling to the corn. Since the density of these $CO_2$ bubbles is less than that of the corn, the total mass of corn plus bubbles is less than an equal volume of liquid. As a result, the corn with bubbles floats to the top. The corn may twist or turn as more or larger bubbles on one side make that side of the kernel less dense.

When the corn kernel reaches the surface, the $CO_2$ bubbles break and join with the air above the liquid. When enough bubbles break, the corn reverts to being denser than the liquid and sinks again. Corn kernels may vary in their reactions because of differences in density due to various factors such as the amount of air or oil inside.

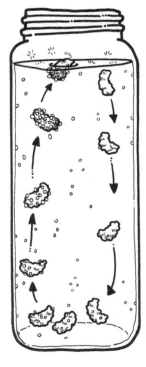

## Management

1. This activity works well with groups of three or four students.
2. Carbonated soft drinks or fizzy-type tablets may be substituted for the seltzer water. Avoid using colored liquids as they limit the observations of the bubbles.
3. For safety, many may prefer to use plastic cups, but jars give the best visual effect.
4. The activity sheets have been designed for flexibility. Use the corn or raisins first; the various seeds in birdseed have varying reactions.

## Procedure

1. Divide into groups and assign roles.
2. Have a whole class discussion of the *Key Question*, "What will happen if we drop corn, raisins, or birdseed into plain or fizzy (carbonated) water?" Share predictions.
3. Invite students to pour water into the jars, no more than 250 mL.
4. Have them drop several corn kernels into the water and observe what happens. Record. Tell them to leave corn in the jar.
5. Break a seltzer tablet into small pieces. Have students add one piece at a time, noting the reaction after each addition. Ask them to observe what happens and record. Tell students to leave corn in the jar.
6. Repeat with raisins and then birdseed.

## Connecting Learning

1. What happens when you place the popcorn or raisins in plain water?
2. What does this tell you about the density of popcorn or raisins compared to the density of water? [They are greater because they sink.]
3. What happens when you add seltzer to the water? [$CO_2$ is produced.]
4. When does a kernel of popcorn rise and sink? Why?
5. Why does the popcorn sometimes twist and turn?
6. Which of the birdseeds sink, float, or move up and down?
7. What are you wondering now?

## Extensions

1. Repeat with other seeds and natural objects. Be sure to have students predict each time.
2. Try it with a paper clip. Guide students to understand that the clip sinks because it is denser than the water, even with carbon dioxide bubbles on it.
3. Have students challenge one another with untried objects.

\* Reprinted with permission from *Principles and Standards for School Mathematics,* 2000 by the National Council of Teachers of Mathematics. All rights reserved.

## Key Question

What will happen if we drop popcorn kernels, raisins, or birdseed into plain or carbonated water?

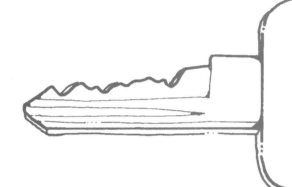

## Learning Goal

### Students will:

observe whether or not there is a change in density to objects in regular and carbonated water.

# Life's Ups & Down's

**Water**

**Choose one:**
raisins or popcorn kernels

Now try mixed birdseed!
Describe what you observe:

Draw a picture.

**Fizzy water and popcorn**
Describe what you observe.

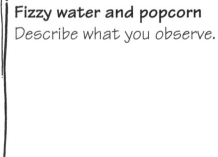

Draw a picture.

# Life's Ups & Downs

Fizzy water
and _____

Draw a picture

Write about it:

## Connecting Learning

1. What happens when you placed the popcorn or raisins in plain water?

2. What does this tell you about the density of the popcorn or raisins compared to the density of water?

3. What happens when you added the seltzer to the water?

4. When does a popcorn kernel rise and sink? Why?

5. Why does the popcorn sometimes twist and turn?

6. Which of the birdseeds sink, float, or move up and down?

7. What are you wondering now?

## Topic
Density and the Cartesian Diver

## Key Question
How can you construct a Cartesian Diver to give the best performance?

## Learning Goal
Students will investigate how changes in the design of a Cartesian Diver will affect its performance and relate this to the concept of density.

## Guiding Documents
*Project 2061 Benchmarks*
- *Organize information in simple tables and graphs and identify relationships they reveal.*
- *Read analog and digital meters on instruments used to make direct measurements of length, volume, weight, elapsed time, rates, and temperature, and choose appropriate units for reporting various magnitudes.*

*NCTM Standards 2000\**
- *Use observations about differences between two or more samples to make conjectures about the populations from which the samples were taken*
- *Solve simple problems involving rates and derived measurements for such attributes as velocity and density*
- *Collect data using observations, surveys, and experiments*

## Math
Measuring
   distance
   time
Averaging
Determining the average rate
Constructing bar graphs

## Science
Physical science
   density
   Cartesian Diver systems

## Integrated Processes
Observing
Collecting and recording data
Interpreting data
Inferring
Applying

## Materials
*For each team:*
   eyedropper
   2-liter soda bottle
   stop watch or watch with a second hand
   metric ruler or tape
   "test tank" such as a liter box

## Background Information
A Cartesian Diver works on the principle that as its density changes it will rise, descend, or remain suspended in water at a certain depth.

Density is determined by the ratio of mass to volume and is defined by the formula: Density = mass/volume. The density of pure water at 4 degrees Celsius is 1 gram/milliliter.

If an eyedropper is used, the density of the Cartesian Diver is a combination of those elements that belong to its system: the rubber bulb, the glass or plastic tube, and the air trapped at the top of the dropper. The water that flows in and out of the diver is **not** part of the diver's system!

After the diver has been placed into the 2-liter bottle, enough water should be added to leave only a very small air space at the top. This will make the diver more responsive. The lid of the bottle must then be closed tightly so that no air escapes.

Squeezing on the sides of the bottle decreases the bottle's volume and increases the pressure inside. This pressure acts equally in all directions throughout the bottle and its contents. The water does not compress but the air compresses readily. The small amount of air at the top of the bottle and the air inside the diver is compressed at the same rate as pressure is applied. This pressure increase causes a flow of water toward the easily compressed part of the diver, its air pocket at the top of the dropper. This leads to a decrease in the diver's volume. Since the volume is the denominator in the formula, a decrease in the volume results in an increase in the diver's density. The density

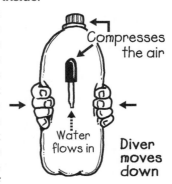

is said to be inversely proportional to the volume. (Note: The water that flows into the diver is not part of its system and therefore is *not* part of its mass. The mass of the diver remains constant. Only its volume changes with changes in pressure.)

If the density is increased sufficiently, the diver will sink. On the other hand, when the pressure is reduced the compressed air in the diver forces out some of the water, increasing the volume of the diver system and thereby reducing its density. If the volume increases sufficiently, the diver will rise to the surface.

An object will float in water if its density is less than that of the water, and it will sink in water if its density is greater than that of water.

Construction of a diver that performs at an optimum level requires that it have the right amount of air space in its system at the outset. Students will discover that a diver that just barely floats can be made to descend rapidly, but will rise slowly. On the other hand, a diver that requires much pressure to descend will descend slowly but rise rapidly. The best performance will come from a diver that balances between these two performances. That is the principle challenge students face in this investigation.

Before using this investigation with students, teachers who have not constructed and experimented with Cartesian Divers are advised to construct one and experiment with different amounts of air in the diver system to gain a feel for what makes a good diver.

Since it is difficult to retrieve a diver from the bottle, it is best to test it in an open tank such as a liter box (or a 2-liter bottle with the top cut off) to make sure it will float properly. It can then be inserted into the bottle for the relays.

Students will observe that as the diver floats at the top of the bottle, the air space in the diver is relatively larger and that when it rests at the bottom it is relatively smaller. From this it follows that the density is less when the air space is larger and the diver will float; and it is greater when the air space is smaller, causing the diver to sink.

There is a difference in the way the Cartesian Diver and a submarine work. Both move up or down by changing their densities. But they do so in different ways: the diver by changing its volume, the submarine by changing its mass.

When a submarine is on the surface, its ballast tanks are full of air. The submarine's mass with air in its tanks is less than the mass of an equivalent volume of water, therefore it is less dense than water and floats. To submerge a submarine, its engineers adjust the buoyancy of the submarine to match that of the water by filling water into the tanks. This equalizes the density of the submarine and the water. They then use the elevators to dive or climb. To remain at the surface, they blow surface air into the tanks to force out the water since using compressed air to empty the tanks is not practical when the submarine is submerged.

## Management
1. Determine which approach to use: discussing the concept of density before beginning the lesson or using the activity to provide an experience with varying densities before discussing the concept.
2. Have students work as teams with each team member keeping a record.

## Procedure
1. Distribute the materials to each team.
2. Permit time for free exploration with the construction of the diver and testing.
3. Have students complete the first recording page after they have selected their best diver design.
4. On *Page 2* have students observe the water levels and the curvature of the bottom of the air bubble at the three levels and draw a picture of each. Then have student teams arrive at a consensus on explaining the behavior of the diver.
5. Next have each team measure the distance traveled in each lap. This is the distance from the bottom of the floating diver to the bottom of the bottle.
6. Have teams share the data in the first column (*distance traveled per lap*) and the third column (*total time for 16 laps*). Require each team to make all the other necessary computations to complete the table. This provides extensive experience computing averages and rates.
7. Have teams complete the bar graphs on *Page 4*.

## Connecting Learning
1. What causes the air in the diver to compress?
2. What causes the diver to ascend and descend?
3. What is the size and shape of the air space when it is at the three levels: floating, suspended midway, or resting at the bottom?
4. How does the operation of the Cartesian Diver compare with that of the submarine?
5. What are you wondering now?

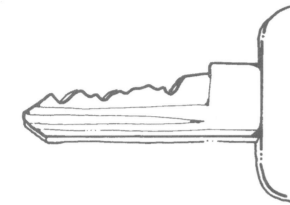

## Key Question

How can you construct a Cartesian Diver to give the best performance?

## Learning Goal

### Students will:

investigate how changes in the design of a Cartesian Diver will affect its performance and relate this to the concept of density.

# Deep Sea Diver Relays

Your assignment is to design a deep sea diver to compete in the Deep Sea Diver Relays.

You will be able to manipulate these variables:
 a. the design of your diver
 b. the technique used in making the dives

The diver is to be made from an eyedropper. The diving tank is a two-liter soft drink container.

Draw a picture of your diver and the diving tank and describe the features of each that you consider important.

Description of diver and its features:

Description of diving tank and its features:

Diver Drawing

Tank Drawing

After making a series of practice dives, describe the technique you will use to control the diver in the Deep Sea Diver Relays.

# Deep Sea Diver RELAYS

After your diver is operational, sketch its appearance showing the water level when the diver is floating, suspended in the center of the tank, and resting at the bottom.

Diver suspended in the center

Diver floating at the top

Diver resting at the bottom

Why does the diver float or sink, rise or descend?

# Deep Sea Diver RELAYS

In the relays you will make eight dives, each consisting of two laps, one down and one up. You will record the total time required for the 16 laps, compute the average time for each lap, and compute the average speed in centimeters per second.

With the diver resting at the top of the tank, measure the distance from the bottom of the diver to the bottom of the tank to determine the distance traveled per lap.

| Name of Contestant | Distance Traveled per lap | Total Distance Traveled in 16 laps | Total Time for 16 laps | Average Time per lap | Average Rate in cm/seconds |
|---|---|---|---|---|---|
| | | | | | |
| | | | | | |
| | | | | | |
| | | | | | |
| | | | | | |
| | | | | | |

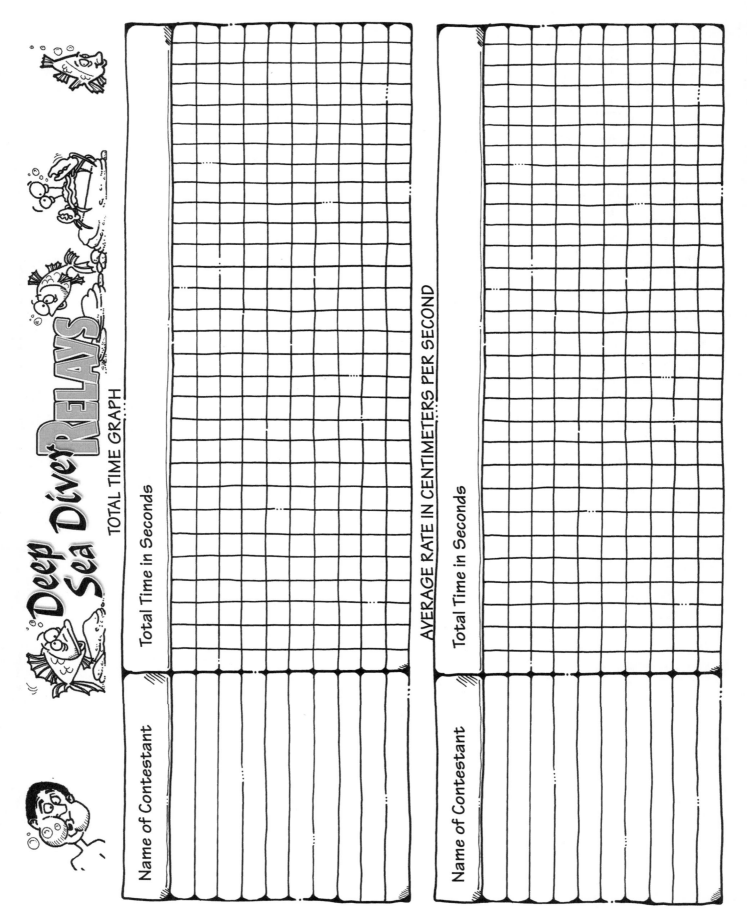

**Deep Sea Diver RELAYS**

TOTAL TIME GRAPH

Total Time in Seconds

Name of Contestant

AVERAGE RATE IN CENTIMETERS PER SECOND

Total Time in Seconds

Name of Contestant

# Deep Sea Diver Relays

## Connecting Learning

1. What causes the air in the diver to compress?

2. What causes the diver to ascend and descend?

3. What is the size and shape of the airspace in the diver when it is at the three levels: floating, suspended midway, or resting at the bottom?

4. How does the operation of the Cartesian Diver compare with that of the submarine?

5. What are you wondering now?

# Some Like It Salty

## Topic
Densities of fresh water and salt water

## Key Question
How can we make a fresh egg float in water?

## Learning Goal
Students will observe a fresh egg in both fresh water and salt water and make inferences as to the water's density.

## Guiding Documents
*Project 2061 Benchmarks*
- *Organize information in simple tables and graphs and identify relationships they reveal.*
- *Offer reasons for their findings and consider reasons suggested by others.*

*NCTM Standards 2000\**
- *Select, create, and use appropriate graphical representations of data, including histograms box plots, and scatterplots*
- *Use observations about differences between two or more samples to make conjectures about the populations from which the samples were taken*
- *Collect data using observations, surveys, and experiments*

## Integrated Processes
Observing
Comparing and contrasting
Controlling variables
Collecting and recording data
Interpreting data
Generalizing

## Math
Measuring
    volume
Graphing
Averaging

## Science
Physical science
    density

## Materials
*For each group of three students:*
    water
    1 fresh egg
    salt
    1 teaspoon
    1 graduated cylinder

## Background Information
Salt water has a greater density than fresh water. By increasing the amount of salt in water, you make it more dense than the egg, causing the egg to float. Students might have noticed that it is easier to float in ocean water than in fresh water or they might be familiar with the fact that scuba divers wear weight belts in ocean water so that they will not float to the surface.

## Management
1. This activity can be done individually or in small groups of three.
2. Students can take turns adding salt and recording.

## Procedure
1. Have students observe a glass of water and an egg. Ask them to predict whether the egg will float or sink. Put the egg into the water. (If the egg is fresh, it will float.)

No Salt          With Salt

2. Ask students what might happen if salt is added to the water. Divide the class into groups to measure and observe what effect the salt in the water will have on the egg.

3. Pour a measured amount of water into the container, 175 mL is recommended. Put the egg into the container of water.
4. Begin to add level teaspoons of salt, a teaspoon at a time, into the water, stirring well after each spoonful until the salt is dissolved.
5. After each teaspoon, observe and record any changes on the student activity sheet.
6. Continue to add salt until the egg breaks the surface of the water.
7. Record on the graph how many teaspoons it took before the egg floated.
8. Add up each team's total and divide the total by the number of teams to find a class average.

## Connecting Learning
1. What happened after the first few teaspoons? ...teaspoons 4 and 5? ...teaspoons 6 and 7? ...teaspoons 8 and 9?
2. Did every group have the same results? Why or why not?
3. How can these results help us to understand why a diver wears weight belts in salt water?
4. Will more salt make the egg float even higher?
5. What results would you expect if we tried different liquids? Explain.
6. What other substances could we add to the water that would make the egg float?

\* Reprinted with permission from *Principles and Standards for School Mathematics*, 2000 by the National Council of Teachers of Mathematics. All rights reserved.

## Key Question

How can we make a fresh egg float in water?

## Learning Goal

### Students will:

observe a fresh egg in both fresh water and salt water and make inferences as to the water's density.

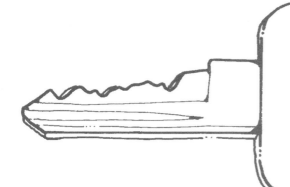

# Some Like it Salty

Pour 175 mL of water into a cup or glass. Gently place the egg into the container. Measure one level teaspoon of salt. Stir the salt into the water completely. Record your observations. Keep adding level teaspoons of salt until the egg floats.

Make a graph to show how many level teaspoons of salt each team used.

Describe what you see after adding each teaspoon.

How many teaspoons of salt did it take to float the egg?

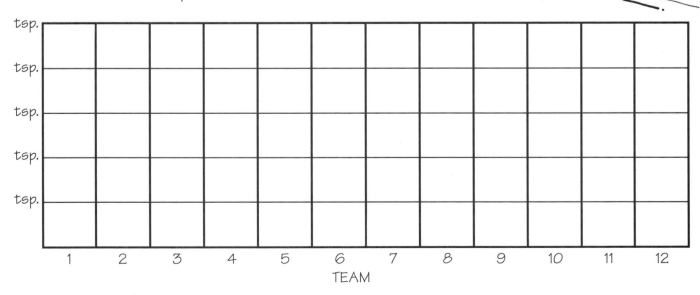

| tsp. | | | | | | | | | | | | |
|------|---|---|---|---|---|---|---|---|---|---|---|---|
| tsp. | | | | | | | | | | | | |
| tsp. | | | | | | | | | | | | |
| tsp. | | | | | | | | | | | | |
| tsp. | | | | | | | | | | | | |
| tsp. | | | | | | | | | | | | |
| | 1 | 2 | 3 | 4 | 5 | 6 | 7 | 8 | 9 | 10 | 11 | 12 |

TEAM

# Some Like It Salty

Our Results:
___mL water
___# tsp. salt


| Group Name | Water (mL) | Number of Teaspoons of Salt |
|---|---|---|
| | | |
| | | |
| | | |
| | | |
| | | |
| | | |
| | | |
| | | |
| | | |
| | | |

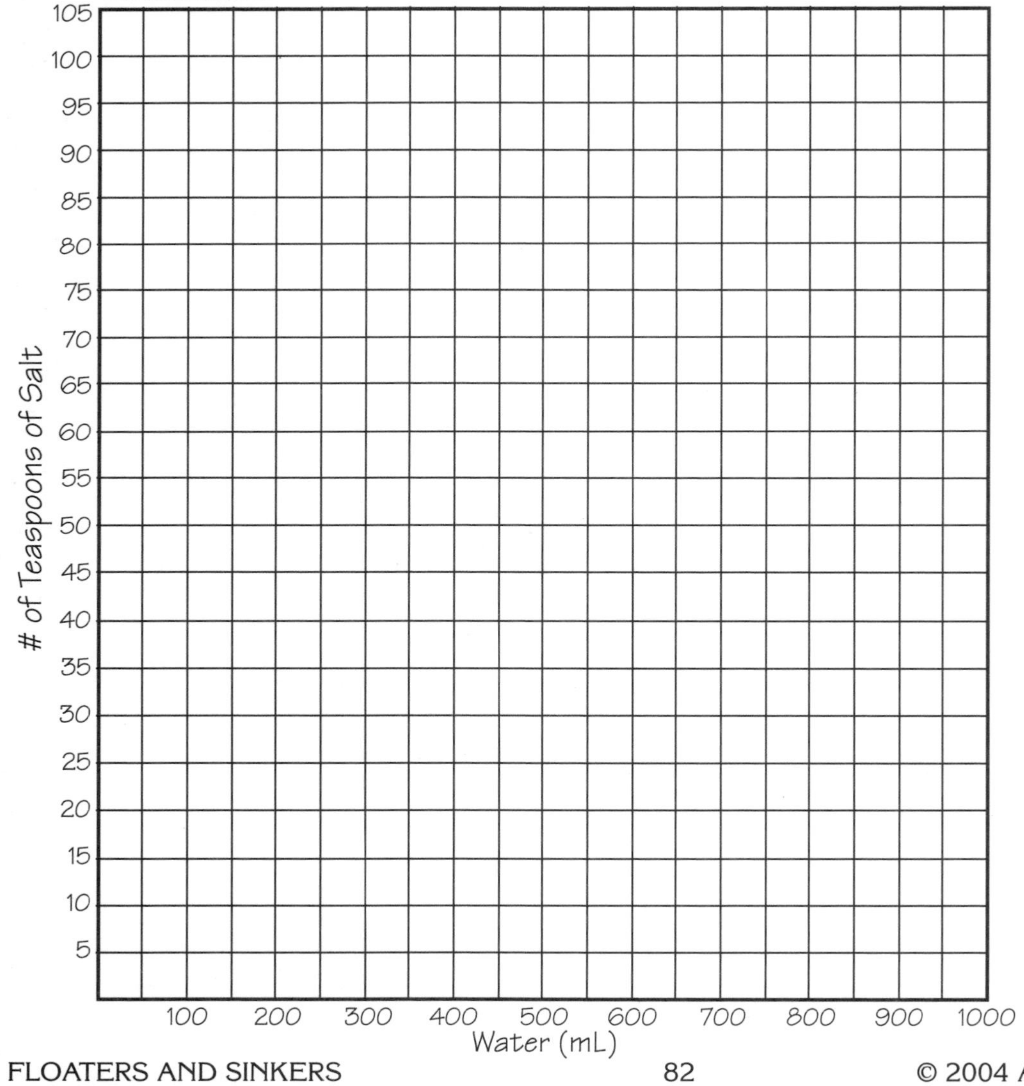

# of Teaspoons of Salt
105
100
95
90
85
80
75
70
65
60
55
50
45
40
35
30
25
20
15
10
5
100 200 300 400 500 600 700 800 900 1000
Water (mL)


In this activity, each group will use a different amount of water in which to float their egg. Gather and record data from at least 10 samples. Graph the results.

What is your conclusion?

 

# Some Like it Salty

## Connecting Learning

1. What happened after the first few teaspoons? ...teaspoons 4 and 5? ...teaspoons 6 and 7? ...teaspoons 8 and 9?

2. Did every group have the same results? Why or why not?

3. How can these results help us to understand why a diver wears weight belts in salt water?

4. Will more salt make the egg float even higher?

5. What results would you expect if we tried different liquids? Explain.

6. What other substances could we add to the water that would make the egg float?

7. What are you wondering now?

## Topic
Volume and density

## Key Question
What is the density of salt?

## Learning Goal
Students will find the mass of various volumes of table salt to determine its density.

## Guiding Documents
*Project 2061 Benchmarks*
- *Organize information in simple tables and graphs and identify relationships they reveal.*
- *Find the mean and median of a set of data.*
- *Decide what degree of precision is adequate and round off the result of calculator operations to enough significant figures to reasonably reflect those of the inputs.*

*NCTM Standard 2000\**
- *Solve simple problems involving rates and derived measurements for such attributes as velocity and density*
- *Collect data using observations, surveys, and experiments*

## Math
Measuring
    volume
    mass
Whole number operations
Graphing
Estimation
    rounding

## Science
Physical science
    density

## Integrated Processes
Observing
Collecting and recording data
Applying
Generalizing

## Materials
Graduated cylinders
300 mL of table salt per group
Balances
Masses
Grid paper

## Background Information
In this activity, students will be finding volume by direct measurement in a graduated cylinder. Although density is usually expressed as grams per cubic centimeter, we are still okay. A milliliter of salt will be approximately equivalent to a cubic centimeter of salt. Therefore just change the unit from milliliters to cubic centimeters when expressing density.

## Management
1. Groups should be from two to five students.
2. Allow about two class periods to complete this investigation, depending upon the amount of equipment available.

## Procedure
1. Have the students find the mass of the empty graduated cylinder and record this data.
2. Direct them to fill the graduated cylinder with 50 mL of salt and find the mass.
3. If necessary, remind the students that they will have to subtract the mass of the empty graduated cylinder to find the mass of the salt.
4. Have the students find the density of the measured volume of salt by dividing mass by volume and rounding off to the nearest hundredth.
5. Direct students to follow the same procedure using other volumes (100, 150, 200…) of salt.
6. When all the data have been collected, have them find the average density of their salt.
7. Distribute the grid paper and ask students to design a graph to display their data.

## Connecting Learning

1. How does the density of salt compare with the density of water?
2. How does the graph you constructed for this activity compare with the graph you made for the mass and volume of water?
3. What can you say about the numbers you recorded in the *Density of Salt* column?
4. What do you think would happen to the density of water if you added salt to it? Explain.
5. What are you wondering now?

## Extensions

1. Have the students find the percent error between their density of salt and the standard density [1.4 g/cm³]. To find percent error:
   a. Find the difference between your density and the standard density.
   b. Divide by the standard density.
   c. Move your decimal point two places to the right (multiply by 100) and add a percent sign.
2. Try this activity again using rock salt. Do the densities change? [because of particle size and space, yes]
3. Have the students graph the mass against the volume of salt used. By connecting these ordered pairs, they should produce a straight line. Can they write an equation for this graph? [Yes, y = 1.4x, or mass = 1.4 volume] What is the slope of this line? [1.4] Can you predict the mass of 300 mL of salt? [300 x 1.4 = 420 g]

\* Reprinted with permission from *Principles and Standards for School Mathematics,* 2000 by the National Council of Teachers of Mathematics. All rights reserved.

# A Salty Problem

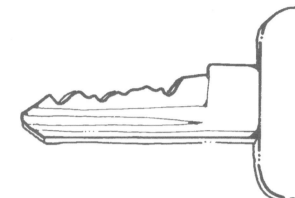

## Key Question

What is the density of salt?

## Learning Goal

### Students will:

find the mass of various volumes of table salt to determine its density.

 # A Salty Problem

1. 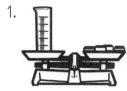 Find the mass of the empty graduated cylinder and record on the chart below.

2.  Measure 50 mL of salt into your graduated cylinder.

3. 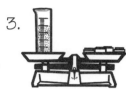 Find the mass of the graduated cylinder with salt in it and record this data.

4. Subtract the two masses to find the mass of salt. Record this under the heading MASS OF SALT.

5. Divide the MASS OF SALT by the VOLUME to find the DENSITY OF SALT and round to the nearest hundredth.

6. Repeat using the indicated volumes of salt.

7. Find the average of your densities.

## Density of Table Salt

| VOLUME | MASS EMPTY | MASS FILLED | MASS OF SALT | DENSITY OF SALT |
|--------|-----------|-------------|--------------|-----------------|
| 50 mL  |           |             |              |                 |
| 100 mL |           |             |              |                 |
| 150 mL |           |             |              |                 |
| 200 mL |           |             |              |                 |
| 250 mL |           |             |              |                 |

AVERAGE DENSITY =

The standard density of table salt is 1.4 g/cm³ (or 1.4 g/mL).

How close did you come?

# A Salty Problem

## Connecting Learning

1. How does the density of salt compare with the density of water?

2. How does the graph you constructed for this activity compare with the graph you made for the mass and volume of water?

3. What can you say about the numbers you recorded in the *Density of Salt* column?

4. What do you think would happen to the density of water if you added salt to it? Explain.

5. What are you wondering now?

## Topic
Relative density of solids and liquids

## Key Question
How will solids and liquids of different densities arrange themselves when combined?

## Learning Goal
Students will measure mass and volume of some liquids to determine relative densities. Various solid objects will be placed into the liquids so that the object's relative densities can be determined.

## Guiding Documents
*Project 2061 Benchmarks*
- *Organize information in simple tables and graphs and identify relationships they reveal.*
- *Equal volumes of different substances usually have different weights.*

*NCTM Standard 2000\**
- *Solve simple problems involving rates and derived measurements for such attributes as velocity and density*
- *Collect data using observations, surveys, and experiments*

## Math
Rational numbers
   decimals
Measuring
   mass
   volume
Equalities and inequalities
   greater than
   less than
   equal to

## Science
Physical science
   density

## Integrated Processes
Observing
Collecting and recording data
Interpreting data
Generalizing

## Materials
*For each group:*
   test tubes or vials—4 small and 1 large
   styrofoam cups—to be used as test tube holders
   corn syrup
   glycerin
   balance
   piece of wood or cork
   piece of pink eraser (Pink Pearl™)
   food coloring, optional
   corn oil
   water
   piece of art gum eraser
   piece of steel alloy (nuts, washer, nails, etc.)
   graduated cylinder

## Background Information
Students are often confused when working with the concept of density. Density is a mass to volume ratio. The puzzlement often arises because of a lack of understanding that dividing one quantity (mass) by a second quantity (volume), can produce a third entirely new quantity (density).

This activity involves the observation of the relative densities of some solids and liquids. Students will first calculate the densities of four liquids and then find how each liquid positions itself in relation to the other liquids when put into a large test tube. The observation of the layering of the liquids along with the calculations should help to reinforce the concept of density. Further reinforcement occurs as solids are added to the liquid layers and students are asked to compare the densities of the solids to the liquids.

The separation of oil and vinegar in salad dressing can be mentioned as an example of the relative densities of liquids. The fact that oil which leaks from ocean tankers or from oil rigs floats on the surface of the ocean is another example. Because it contains salt, sea water is more dense than fresh water. The Great Salt Lake in Utah and the Salton Sea in southern California are very salty and much more dense than sea water. If you were to go swimming in these waters, you would immediately notice how well you can float.

## Management
1. The liquids should be in four containers marked A, B, C, and D. The correct identity of the liquids should be masked.

2. The corn syrup may be used either as is or colored with food coloring. Use of food coloring makes the activity more spectacular. If you do use food coloring, it should be added to the corn syrup a couple of days in advance with the bottle being turned every so often.
3. Food coloring should be added to the water as the water and glycerin are both colorless.
4. Inverted styrofoam cups with holes in the bottom make good test tube holders.
5. Density is measured in grams per milliliter on the activity sheet. If you do not have equipment for measuring liquids in milliliters, 1 mL of water has a mass of about 1 gram. Find the mass of an empty test tube, add enough water to add 10 grams to the mass of that test tube, mark the test tube at the water level. All test tubes or vials should be marked at this same point so that students will know how much of each liquid should be poured into them.
6. Be sure that liquids from all four small test tubes and the solids can be contained easily in the large test tube.

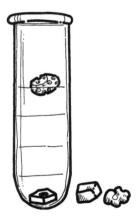

7. If students find it difficult to find the mass of the test tubes once the liquid is in them, you might have them find the mass of the test tube in the styrofoam cup holder both empty and filled.
6. Be sure that the pieces of solids are small enough that they can fit into the test tube easily and can pass each other no matter what order the students choose to drop them into the test tube.

7. The liquid layers may be left to display for several days. After several days the two middle layers, the glycerin and the water, may start to mix. If the class wonders what will happen if you stir everything and then let it settle, the bottom three layers will mix and become one. Only the corn oil will remain a separate layer.

## Procedure
1. Divide the class into groups of four to six students.
2. Direct students to label their four small test tubes A, B, C, and D and to find the mass of each of them empty.
3. Have the students pour 10 mL of liquid A into test tube A. Do the same for test tubes B, C, and D with the respective liquids.
4. Direct the students to find and record the mass of each of the test tubes with the liquids.
5. Review the calculations for finding density that are given in the table. Have the students find the density of each liquid.
6. Distribute a large test tube to each group. Direct the students to pour each liquid into the large test tube, beginning with the one of greatest density and continuing in order until all liquids are in the large test tube.
7. Ask the students to determine which solid is to be No. 1, No. 2, etc. Have them list the numbers and solids on the activity sheet. Direct them to drop the solids into the liquids in varied orders and record their observations.

## Connecting Learning
1. What was the volume of each of the liquids? [10 mL] If all of them had the same volume, why are they layered? [They have different masses.]
2. What do you know about the liquid that went to the bottom of the test tube? [Its mass per volume was greater than the liquids above it.]
3. Does it make a difference which of the solids is dropped in first?
4. What will happen if everything is mixed up and allowed to settle?
5. Do you know the identity of any of the liquids? [perhaps the water] How do you know? [its density is one]
6. What other liquids would you like to test? ...what solids?

## Extension
Select five other liquids and find their relative densities.

* Reprinted with permission from *Principles and Standards for School Mathematics,* 2000 by the National Council of Teachers of Mathematics. All rights reserved.

# DENSER Sensor

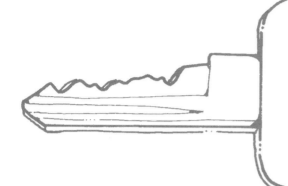

## Key Question

How will solids and liquids of different densities arrange themselves when combined?

## Learning Goal

measure mass and volume of some liquids to determine relative densities. Various solid objects will be placed into the liquids so that the objects' relative densities can be determined.

# DENSER Sensor

Label the four (4) small test tubes A, B, C, and D.
Find the mass of each one that is empty.
Pour 10 mL of liquid into each
Find the mass of the full test tubes, subtract,
and compute the density of each liquid.

| | Mass of Empty Test Tube (g) | Mass of Test Tube with Liquid (g) | Mass of Liquid (g) | Volume of Liquid (mL) | Density of Liquid Mass/Volume (g/mL) |
|---|---|---|---|---|---|
| A | | | | | |
| B | | | | | |
| C | | | | | |
| D | | | | | |

Pour each liquid in order of density into the large container. Begin with the liquid with the greatest density. Record your observations.

Identify the solids: 1 _____ 2 _____ 3 _____ 4 _____

- Drop the solids (one at a time) into the container with the liquids.
- Allow them to come to rest. Compare the solids to the liquids by completing the following statements:

The density of solid #1 is greater than ___ and less than ___.

The density of solid #2 is greater than ___ and less than ___.

The density of solid #3 is greater than ___ and less than ___.

The density of solid #4 is greater than ___ and less than ___.

# DENSER Sensor

Draw and label the solids and liquids after you have poured or dropped them into the container.

**solids**

**liquids**

# DENSER Sensor

## Connecting Learning

1. What was the volume of each of the liquids? If all of them had the same volume, why are they layered?

2. What do you know about the liquid that went to the bottom of the test tube?

3. Does it make a difference which of the solids is dropped in first? Explain.

4. What will happen if everything is mixed up and allowed to settle?

5. Do you know the identity of any of the liquids? How do you know?

6. What other liquids would you like to test? ...what solids?

# See Level

## Topic
Volume and density

## Key Question
Which liquid will float on which?

## Learning Goal
Students will find the densities of five different liquids and predict the layering in a flotation column.

## Guiding Documents
*Project 2061 Benchmarks*
- *Equal volumes of different substances usually have different weights.*
- *Organize information in simple tables and graphs and identify relationships they reveal.*

*NCTM Standards 2000\**
- *Solve simple problems involving rates and derived measurements for such attributes as velocity and density*
- *Work flexibly with fractions, decimals, and percents to solve problems*
- *Collect data using observations, surveys, and experiments*

## Math
Measuring
 volume
 mass
Rational numbers
 decimals
Averaging

## Science
Physical science
 density
 specific gravity

## Integrated Processes
Observing
Comparing and contrasting
Predicting
Collecting and recording data
Interpreting data
Applying
Generalizing

## Materials
Graduated cylinders or beakers
Balance
Masses
Motor oil
Vegetable oil
Rubbing alcohol
Salt
Paper towels

## Background Information
To calculate the density of a liquid, divide the mass of the liquid sample by the volume of the sample. The following is a list of specific gravities (the ratio of the density of a liquid to the density of water) obtained from the investigation. These numbers are for comparison only, your calculations may vary somewhat.

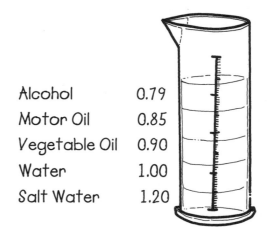

| | |
|---|---|
| Alcohol | 0.79 |
| Motor Oil | 0.85 |
| Vegetable Oil | 0.90 |
| Water | 1.00 |
| Salt Water | 1.20 |

To make the salt water solution, use about 200 mL of salt for each liter of water. Stir well and allow to set overnight to become as saturated as possible.

When pouring liquids to form a flotation column, avoid mixing of liquids by tilting a large graduated cylinder and very gently pouring the liquids down the side. If you are careful, the liquids are very cooperative about wedging themselves in where they belong. You may wish to add some food coloring to the salt solution for visibility.

## Management

1. Have a large quantity of paper towels available to clean out beakers and accidental spills.
2. Have students do measurements of the oil last; water, salt water, and alcohol are easy to clean out of beakers, oil is not.
3. If triple-beam balances are available, use them. Other balances can be used, but they are less accurate. To obtain greater accuracy with balances that will not measure less than a gram, use a greater volume of liquid. Liquid samples in these cases should be at least 150 mL.
4. This investigation requires several class periods if class time is given to do the calculations.

## Procedure

1. Ask the *Key Question: Which liquid will float on which?* Tell students that they are to determine the order of the liquids by finding their densities.
2. Have students find the mass of a container (clean beaker or graduated cylinder).
3. Have them obtain a sample of liquid from the stock supply (50 to 200 mL).
4. Tell students to determine the volume of the liquid.
5. Direct them to determine the mass of the container with the liquid.
6. Have students repeat steps 3, 4, and 5 twice with different liquids.
7. Ask students to calculate the mass of each sample of liquid by subtracting the mass of the container from the mass of the container and liquid.
8. Direct them to calculate the density of each sample of liquid (density = mass/volume) and then find the average density for each type.
9. Have students order the liquids from most dense to least dense by number the liquids on the student sheet. Invite them to carefully pour the liquids in a graduated cylinder or other narrow cylindrical container to check whether their calculations hold true in the real world.

## Connecting Learning

1. Which liquids do you think would float on which if they were all poured into the same beaker? After predictions are made, make a column to test the predictions. In the process of discussion, be sure that the students recognize that the numbers they obtained as a quotient in the density calculation is the mass of 1 mL of the liquid.
2. Why do you think someone would want to know the densities of liquids? [To predict their reaction to another liquid, usually water. The ratio of the density of a liquid to the density of water is called *specific gravity*.]
3. How much variance did you have in the densities of the three trials of each liquid? How do you account for these differences?
4. What are you wondering now?

\* Reprinted with permission from *Principles and Standards for School Mathematics,* 2000 by the National Council of Teachers of Mathematics. All rights reserved.

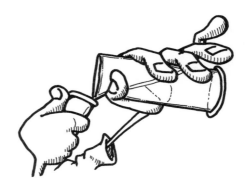

# SEE Level

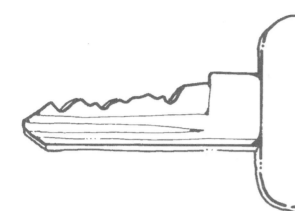

## Key Question

Which liquid will float on which?

## Learning Goal

### Students will:

find the densities of five different liquids and predict the layering in a flotation column.

Slow...

# SEE Level

Slow...

| Liquid | Sample Number | Mass of Container | Mass of Container + Liquid | Mass of Liquid | Volume | Density | Average Density |
|--------|---------------|-------------------|----------------------------|----------------|--------|---------|-----------------|
| Water | 1 | | | | | | |
| | 2 | | | | | | |
| | 3 | | | | | | |
| | 1 | | | | | | |
| | 2 | | | | | | |
| | 3 | | | | | | |
| | 1 | | | | | | |
| | 2 | | | | | | |
| | 3 | | | | | | |
| | 1 | | | | | | |
| | 2 | | | | | | |
| | 3 | | | | | | |
| | 1 | | | | | | |
| | 2 | | | | | | |
| | 3 | | | | | | |

# See Level

## Connecting Learning

1. Which liquids do you think would float on which if they were all poured into the same beaker?

2. Why do you think someone would want to know the densities of liquids?

3. How much variance did you have in the densities of the three trials of each liquid? How do you account for these differences?

4. What are you wondering now?

## Topic
Density

## Key Question
Which sample will have the greatest density?

## Learning Goal
Students will compare the density of different sizes of the same kind of wood.

## Guiding Documents
*Project 2061 Benchmarks*
- *Graphs can show a variety of possible relationships between two variables. As one variable increases uniformly, the other may do one of the following: always keep the same proportion to the first, increase or decrease steadily, increase or decrease faster and faster, get closer and closer to some limiting value, reach some intermediate maximum or minimum, alternately increase and decrease indefinitely, increase and decrease in steps, or do something different from any of these.*
- *Calculate the circumferences and areas of rectangles, triangles, and circles, and the volumes of rectangular solids.*
- *Organize information in simple tables and graphs and identify relationships they reveal.*
- *The choice of materials for a job depends on their properties and how they interact with other materials. Similarly, the usefulness of some manufactured parts of an object depends on how well they fit together with the other parts.*

*NCTM Standards 2000\**
- *Work flexibly with fractions, decimals, and percents to solve problems*
- *Collect data using observations, surveys, and experiments*
- *Solve simple problems involving rates and derived measurements for such attributes as velocity and density*

## Math
Measuring
  length
  mass
Using formulae
  volume
  density
Graphing

## Science
Physical science
  density

## Integrated Processes
Observing
Collecting and recording data
Interpreting data
Generalizing
Applying

## Materials
Balance
Masses
Metric rulers
Various-sized rectangular pieces of the same kind of wood
Graph paper

## Background Information
The physical properties of wood are extremely diverse, not only between different species, but within species and even within the parts of an individual tree. Wood density is determined using oven-dried wood. The main factor governing the density of a given wood sample is the ratio of cell-wall to cell-cavity volume.

For this investigation, it is suggested that pine be used because of the ready availability. Pieces of various lengths and thicknesses can be cut from a single board.

## Management
1. Wood samples can be obtained from wood shops at school, lumber yards, and cabinet shops.
2. Have students work in small groups of two or three.
3. This investigation should take about two periods of 45 minutes.
4. The wood samples should be numbered or lettered.

## Procedure

1. Don't tell the students that the wood is all the same kind.
2. Instruct students to measure the length, width, and height of the blocks of wood in centimeters and record.
3. Ask them to calculate the volume of the sample using the formula $V = l \times w \times h$.
4. Have the students find the mass of each sample and record.
5. Direct students to find the density of each sample by using the formula $D = m/v$.
6. Distribute graph paper and ask students to graph their data with *mass* on the vertical axis and *volume* on the horizontal axis.

## Connecting Learning

1. How did the densities compare? (The densities should be very close.)
2. If these pieces of wood were all cut off the same piece of wood, why would the densities differ? [Knots and center wood are more compacted.]
3. What does your graph tell you about the density of this wood? [When the plotted data points are connected with a best-fit line, the line represents the density of the wood. Because the points are located very close to the best-fit line, it indicates that the pieces of wood are of the same type.]
4. What are you wondering now?

## Extension

Try to obtain a piece of wood just cut, but not dried. Compare the density of it to the density of lumberyard wood. Is there a difference? Why? [There is water in the fresh wood.]

\* Reprinted with permission from *Principles and Standards for School Mathematics,* 2000 by the National Council of Teachers of Mathematics. All rights reserved.

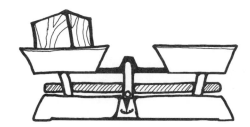

## Key Question

Which sample will have the greatest density?

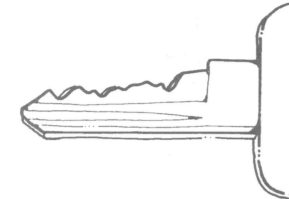

## Learning Goal

### Students will:

compare the density of different sizes of the same kind of wood.

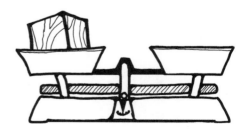

## Predictions:

Which sample will have the greatest density?

Which sample will have the least density?

Explain the reasons for your choices.

**Record the data from your samples on the chart below and then calculate the volume and density.**

| Sample | Length (cm) | Width (cm) | Height (cm) | Volume (cm³) | Mass (grams) | $D = \frac{m}{v}$ |
|--------|-------------|------------|-------------|--------------|--------------|-------------------|
|        |             |            |             |              |              |                   |
|        |             |            |             |              |              |                   |
|        |             |            |             |              |              |                   |
|        |             |            |             |              |              |                   |
|        |             |            |             |              |              |                   |

Using graph paper, graph the mass and volume for each sample above.

# Connecting Learning

1. How did the densities compare?

2. If these pieces of wood were all cut off the same piece of wood, why would the densities differ?

3. What does your graph tell you about the density of this wood?

4. What are you wondering now?

## Topic
Density of different types of wood

## Key Question
Why are some woods better for certain purposes than other woods?

## Learning Goal
Students will compute the volume and density of different kinds of woods.

## Guiding Documents
*Project 2061 Benchmarks*
- *Calculate the circumferences and areas of rectangles, triangles, and circles, and the volumes of rectangular solids.*
- *Organize information in simple tables and graphs and identify relationships they reveal.*

*NCTM Standards 2000\**
- *Work flexibly with fractions, decimals, and percents to solve problems*
- *Collect data using observations, surveys, and experiments*
- *Solve simple problems involving rates and derived measurements for such attributes as velocity and density*

## Math
Measuring
   length
   mass
Calculating volume
Graphing
Estimating

## Science
Physical science
   density

## Integrated Processes
Observing
Classifying
Collecting and recording data
Interpreting data
Generalizing
Applying

## Materials
Balsa wood plane
Stick
Balance
Masses
Metric rulers
Similarly shaped rectangular pieces of various woods

## Background Information
Wood densities are calculated using oven-dry wood samples. Following are some densities for various species of wood found in *American Institute of Physics Handbook*, Dwight E. Gray, Coordinating Editor, McGraw-Hill Book Co., Inc. 1957.

| Common Name | Density (g/cm³) |
| --- | --- |
| Balsa, tropical American | 0.122-0.20 |
| Birch, paper | 0.600 |
| Birch, yellow | 0.668 |
| Cedar, eastern red | 0.492 |
| Cherry, wild red | 0.425 |
| Chestnut | 0.454 |
| Cottonwood, eastern | 0.433 |
| Dogwood, flowering | 0.796 |
| Elm, American | 0.554 |
| Fir, balsam | 0.414 |
| Ironwood, black | 1.077 |
| Locust, honey | 0.666 |
| Maple, silver | 0.506 |
| Maple, sugar | 0.676 |
| Oak, red | 0.657 |
| Oak, white | 0.710 |
| Pine, eastern white | 0.373 |
| Walnut, black | 0.562 |

## Management
1. Working with small groups of two or three.
2. This investigation should take about two or three periods of 45 minutes.
3. Wood samples may be obtained from school wood shops, lumber yards, and cabinet shops.
4. Know what types of wood are in your samples.
5. Save several "mystery" wood samples (unidentified) for distribution at the end of the lesson or to be used as an assessment.

## Procedure

1. Throw a balsa wood plane and then a stick.
2. Ask the *Key Question*.
3. Tell the students that they will be investigating one of the reasons some woods are used for special purposes.
4. Direct them to measure (cm) and record the length, width, and height of samples of wood.
5. Have the students calculate the volume of the same samples using the formula $V = l \times w \times h$.
6. Have them find the mass of each sample and record.
7. Direct the students to find the density of each sample by using the formula $D = m/v$.
8. Make a density chart for the various types of wood.
9. Distribute mystery samples and ask students to identify the types.

## Connecting Learning

1. Which woods were the most dense? ...the least dense?
2. In construction, which woods do you think are used more for structural jobs? Explain.
3. What type of wood is the plane made of? Why that type?
4. What other differences do you see in the types of wood?
5. What are you wondering now?

\* Reprinted with permission from *Principles and Standards for School Mathematics,* 2000 by the National Council of Teachers of Mathematics. All rights reserved.

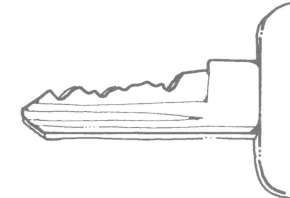

## Key Question

Why are some woods better for certain purposes than other woods?

## Learning Goal

compute the volume and density of different kinds of woods.

# Are You As DENSE As I Am Again?

**Why are some woods better for certain purposes than other woods?**

Predictions:

Which wood sample is least dense?

Which wood sample is most dense?

Record your sample information on the chart below. Calculate the volume and density.

| Wood Type | Length (cm) | Width (cm) | Height (cm) | Volume (cm³) | Mass (grams) | $D = \frac{m}{v}$ |
|-----------|-------------|------------|-------------|--------------|--------------|-------------------|
|           |             |            |             |              |              |                   |
|           |             |            |             |              |              |                   |
|           |             |            |             |              |              |                   |
|           |             |            |             |              |              |                   |
|           |             |            |             |              |              |                   |

| Wood | 0.1 | 0.3 | 0.5 | 0.7 | 0.9 | 1.1 | 1.3 | 1.5 | 1.7 |
|------|-----|-----|-----|-----|-----|-----|-----|-----|-----|
|      |     |     |     |     |     |     |     |     |     |
|      |     |     |     |     |     |     |     |     |     |
|      |     |     |     |     |     |     |     |     |     |
|      |     |     |     |     |     |     |     |     |     |
|      |     |     |     |     |     |     |     |     |     |
|      |     |     |     |     |     |     |     |     |     |

My Mystery Wood is

# Are You As DENSE As I Am Again?

## Connecting Learning

1. Which woods are the most dense? ...the least dense?

2. In construction, which woods do you think are used more for structural jobs? Explain.

3. What type of wood is the plane made of? Why that type?

4. What other differences do you see in the types of wood?

5. What are you wondering now?

# Floating Wood

**Topic**
Density of wood and water

**Key Question**
What part of a block would be submerged if placed in water?

**Learning Goal**
Students will compare the density of wood to the density of water to determine how much of the wood would be submerged in the water.

**Guiding Documents**
*Project 2061 Benchmarks*
- *The expression a/b can mean different things: a parts of size 1/b each, a divided by b, or a compared to b.*
- *Calculate the circumferences and areas of rectangles, triangles, and circles, and the volumes of rectangular solids.*

*NCTM Standards 2000\**
- *Select and apply techniques and tools to accurately find length, area, volume, and angle measures to appropriate levels of precision*
- *Understand relationships among the angles, side lengths, perimeters, areas, and volumes of similar objects*
- *Work flexibly with fractions, decimals, and percents to solve problems*
- *Collect data using observations, surveys, and experiments*
- *Solve simple problems involving rates and derived measurements for such attributes as velocity and density*

**Math**
Measuring
    length
    mass
Estimating
    rounding
Whole number operations
Rational numbers
    decimals
Averaging
Using formulae
    volume
    density

**Science**
Physical science
    density

**Integrated Processes**
Observing
Predicting
Collecting and recording data
Interpreting data
Generalizing
Applying

**Materials**
Dry pine blocks of various sizes
Metric ruler
Balance
Masses
Tub of water to float wood in
Paper towels

**Background Information**
Water has the density of 1 g/cm$^3$ at 4°C. Anything with a density less than 1 will float. Anything with a density greater than 1 will sink. By computing the density of the pine block, you can find the portion that would be submerged. Example: If the density of the wood is .5 g/cm$^3$, then 1/2 of the block would be submerged. If the density is .3 g/cm$^3$, then about 1/3 would be submerged.

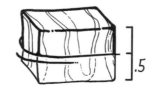 

## Management

1. The suggested time is thirty minutes.
2. This activity works well with groups of three.
3. Make sure that the tubs for floating the wood are large enough to accommodate the largest block of wood.
4. Try to have a block that will not float horizontally but diagonally to show that the part submerged is still .5 or whatever the density is.

## Procedure

1. Place several "oceans" around the room with a supply of towels nearby.
2. Let each group send a representative to choose a block of wood.
3. Direct the students to measure and find the average data for length, width, and height.
4. Have them find the mass in grams and compute the density by dividing the mass by the volume.
5. Using the density information, have students mark a prediction line for flotation.
6. Direct them to check their prediction.

## Connecting Learning

1. How close was your prediction?
2. What explanation do you have for prediction lines that were off?
3. Why did you have to divide by 1000 to find the volume measurement?
4. Why does the density of your block tell you where the float line will be? [It is easy to predict because water has a density of 1.]
5. What is the mass of a cubic meter of dry pine wood (or whatever type of wood you used) based upon this investigation?
6. What is the approximate mass of a truckload of dry pine wood, closely stacked, if the truck bed that is covered measured 2.5 meters by 4 meters and the stack is 1.7 meters high (all measured to the nearest tenth of a meter)?

## Extension

See *Afloat, Ship Shape,* and *Clay Boats* in this book.

\* Reprinted with permission from *Principles and Standards for School Mathematics,* 2000 by the National Council of Teachers of Mathematics. All rights reserved.

Floating Wood

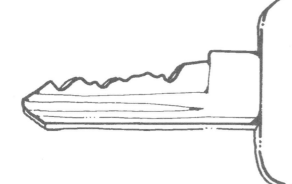

## Key Question

What part of a block would be submerged if placed in water?

## Learning Goal

### Students will:

compare the density of wood to the density of water to determine how much of the wood would be submerged in the water.

# Floating Wood

## Can You Predict the Flotation Line For a Block of Dry Pine?

1. Make four measures, each to the nearest millimeter, of each dimension of your boat: length, width, and height. Measure along the four parallel edges in each case.

2. Find the average of each set of measures and enter into the table.

3. Compute the volume using the average measures.

4. Change this volume from mm³ to cm³ (divide by 1000).

5. Round to the nearest tenth of a cubic centimeter.

| | Length | Width | Height |
|---|---|---|---|
| First Edge | mm | mm | mm |
| Second Edge | mm | mm | mm |
| Third Edge | mm | mm | mm |
| Fourth Edge | mm | mm | mm |
| Average | mm | mm | mm |

Volume = lwh ÷ 1000 = _____ cm³

Volume of block _____ cm³

Mass of block _____ grams

Density of block _____ grams/cm³

I predict the water line will be _____ **mm** from the bottom.

Lightly mark your block where you predict the water line will be if placed in water. Test to see how accurately you predicted the results.

## Connecting Learning

1. How close was your prediction?

2. What explanation do you have for prediction lines that were off?

3. Why did you have to divide by 1000 to find the volume measurement?

4. Why does the density of your block tell you where the float line will be?

5. What is the mass of a cubic meter of dry pine wood (or whatever type of wood you used) based upon this investigation?

6. What is the approximate mass of a truckload of dry pine wood, closely stacked, if the truck bed that is covered measured 2.5 meters by 4 meters and the stack is 1.7 meters high (all measured to the nearest tenth of a meter)?

# Topic
Flotation of wood in tap water and salt water

# Key Question
What happens when a ship goes from fresh water into salt water?

# Learning Goals
Students will:
1. compare the density of a block of wood to that of tap water and salt water, and
2. they will determine the flotation line of the wood in both situations.

# Guiding Documents
*Project 2061 Benchmarks*
• *Mathematics is helpful in almost every kind of human endeavor—from laying bricks to prescribing medicine or drawing a face. In particular, mathematics has contributed to progress in science and technology for thousands of years and still continues to do so.*
• *Use, interpret, and compare numbers in several equivalent forms such as integers, fractions, decimals, and percents.*

*NCTM Standards 2000\**
• *Develop strategies to determine the surface area and volume of selected prisms, pyramids, and cylinders*
• *Work flexibly with fractions, decimals, and percents to solve problems*
• *Collect data using observations, surveys, and experiments*
• *Solve simple problems involving rates and derived measurements for such attributes as velocity and density*

# Math
Measuring
 length
Estimating
 rounding
Averaging
Using formulae
 volume
Rational numbers
 ratios
 decimals

# Science
Physical science
 density

# Integrated Processes
Observing
Predicting
Collecting and recording data
Interpreting data
Generalizing
Applying

# Materials
Dry rectangular blocks of wood
Metric rulers
Balance
Masses
Tub of water to float wood in
Salt
Pill bottles or small vials

# Background Information
Tap water has the density of 1 g/cm³ at 4°C. Salt water has the density of 1.2 g/cm³ if prepared the following way:

Add salt a little at a time to your tap water, stirring to dissolve. Stop when you see that the salt is no longer dissolving.

An object will float higher in salt water than in tap water. By comparing the density of the wood to the water it will be floated in, you can figure how much of the wood will be in the water. The ratio of the densities will give you what portion of the block will be under water.

Example:  The density of wood is .6 g/cm³
 The density of tap water is 1 g/cm³
 The ratio density of wood is .6
 The density of water is 1.0

Therefore, 6/10 of the wood would be under water. Mark a line 6/10 of the way and try it!

$$\frac{\text{The ratio density of wood}}{\text{density of salt water}} = \frac{.6}{1.2} = .5$$

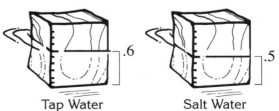

Tap Water          Salt Water

Therefore, 5/10 or 1/2 of the wood is under water. Mark a line and check.

## Management

1. Students should work in pairs.
2. This activity can be done in one class period or two, depending on the skill of the students and equipment available.
3. Students should have done the investigations *Wat-ar Densities* and *A Salty Solution* to know that the densities of fresh water and salt water are different.
4. Have the students float their wooden ships in tap water and then change the tap water to salt water.
5. When determining the densities of the water and salt water, students should use samples from the water they will actually be using. They will need to complete all the tap water portion of the activity, add salt to the tap water, find its density, then finish the salt water portion.
6. Tell the students to remove the wood as soon as possible from the tap water to avoid absorption of water causing a change in mass.

## Procedure

1. Direct the students to take a sample of their tap water and find its density using the pill bottle sample.
2. Have the students find the mass, volume, and density of their block of wood.
3. Direct them to find the ratio of the wood's density to the water's density.
4. Have them mark a line showing where it will float in tap water.
5. Allow time for students to float the block to check their accuracy and make any line corrections necessary. If their prediction lines were correct, fine. If not, have them mark the correct place with another color.
6. Repeat steps 1-4 using salt water.

## Connecting Learning

1. What happened to the waterline from tap water to salt water?
2. What implications would this have in the real world? [Ships going from salt water to fresh water would ride lower in the water. This would need to be considered when determining how full to load the ship.]
3. What would ships have to worry about most—coming from salt water into a port of fresh water or leaving a fresh water port and sailing into salt water (temperature of the water remains the same)? [Coming into port from salt water to fresh water. They will sink further into the water. They need to make sure their cargo isn't riding so low in the water that their ship hits bottom.]
4. What are you wondering now?

\*   Reprinted with permission from *Principles and Standards for School Mathematics,* 2000 by the National Council of Teachers of Mathematics. All rights reserved.

# Afloat

**Key Question**

What happens when a ship goes from fresh water into salt water?

## Learning Goal

**Students will:**

1. compare the density of a block of wood to that of tap water and salt water, and
2. determine the flotation line of the wood in both situations.

# Afloat

Use an empty pill bottle to find the densities of the oceans.

1. To find the volume of the pill bottle:
   a. Measure the diameter.
   b. Find 1/2 the diameter= radius (r).
   c. Square the radius (r x r).
   d. Multipy by 3.14 ($\pi$).
   e. Multiply by the height.
2. Find the mass of the empty bottle.
3. Fill the bottle with water and find the mass.
4. The difference between full and empty is the mass of the water.
5. Divide the mass by the volume—**Density!**

| Ocean | 3.14 • r • r • height = | Volume | Mass Of Bottle Empty | Mass Of Bottle Full | Ocean Mass | Density |
|-------|-------------------------|--------|----------------------|---------------------|------------|---------|
| Tap   |                         |        |                      |                     |            |         |
| Salt  |                         |        |                      |                     |            |         |

Now, choose a wooden boat. To calculate its density, do the following:

1. Find the mass of the boat.
2. Measure the length, width and height.
3. Multiply to find the volume.
4. Divide the mass by the volume—**Density!**

| My Boat is Block #__ | Mass | Length | Width | Height | Volume | Density |
|----------------------|------|--------|-------|--------|--------|---------|
|                      |      |        |       |        |        |         |

Densities of:

1. The tap water ocean _____ g/cm³
2. The salt water ocean _____ g/cm³
3. The wooden boat _____ g/cm³

Remember: Density = $\frac{\text{mass (grams)}}{\text{volume (cm}^3)}$

Compare the density of your wooden boat and the tap water ocean. Mark a line to show where you think your boat will float. Test it.

What will happen to the water line if you place your wooden boat into the salt water ocean?

Mark where the float line will be when placed in the salt water and test it.

Describe what you just noticed. Do cargo ships have this problem?
Explain what they must do to adjust for this?

# Afloat

## Connecting Learning

1. What happened to the water line from tap water to salt water?

2. What implications would this have in the real world?

3. What would ships have to worry about most—coming from salt water into a port of fresh water or leaving a fresh water port and sailing into salt water (temperature of the water remains the same)?

4. What are you wondering now?

## Topic
Density and buoyancy

## Key Questions
1. How can you make 30 grams of clay float?
2. What design will allow you to carry the most cargo?

## Learning Goal
Students will design clay boats to discover which shape can carry the most cargo.

## Guiding Documents
*Project 2061 Benchmarks*
- *Be aware that there may be more than one good way to interpret a given set of findings.*
- *Measuring instruments can be used to gather accurate information for making scientific comparisons of objects and events and for designing and constructing things that will work properly.*

*NCTM Standard 2000\**
- *Collect data using observations, surveys, and experiments*

## Math
Measuring
    mass
Estimating
Graphing

## Science
Physical science
    density
    buoyancy

## Integrated Processes
Observing
Predicting
Collecting and recording data
Generalizing
Applying

## Materials
Plasticene clay
Containers of water for floating boats
Uniform masses (tiles, centicubes, marbles, pennies,
    paper clips, etc.)
Paper towels

## Background Information
Archimede's Principle says that any object wholly or partially immersed in a liquid is buoyed up by a force equal to the weight of the liquid displaced.

In this investigation, students will discover that their boat needs sides and that a cup-like shape often works best.

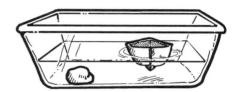

## Management
1. Groups of two or three work well for this activity.
2. Allow about 45 minutes with time for free exploration with the clay at the beginning.
3. Encourage the groups to name their vessels.
4. You will need to use a clay that does not deteriorate in water. If Plasticene clay cannot be found, try using florist's clay.

## Procedure
1. Give each group 30 grams of clay.
2. Have them make shapes using all 30 grams until they find a shape that floats. Direct them to sketch the boat and find its mass.
3. Instruct the students to put the boat in the water and add masses until it sinks.
4. Give them three tries to manipulate their designs with the goal of holding more masses. The boat design that holds the most after three tries should be used for comparisons with other groups.

## Connecting Learning
1. For the designs that floated, which had more mass: the boat or the cargo?
2. For those that sank, which had more mass?
3. Is there a critical weight factor? Explain.
4. What other factors may have determined whether the boats would float?
5. What are you wondering now?

\* Reprinted with permission from *Principles and Standards for School Mathematics,* 2000 by the National Council of Teachers of Mathematics. All rights reserved.

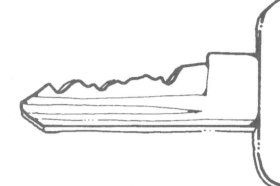

## Key Questions

1. How can you make 30 grams of clay float?
2. What design will allow you to carry the most cargo?

## Learning Goal

design clay boats to discover which shape can carry the most cargo.

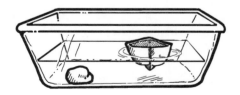

# CLAY BOATS

Will it float?  How much will it hold?

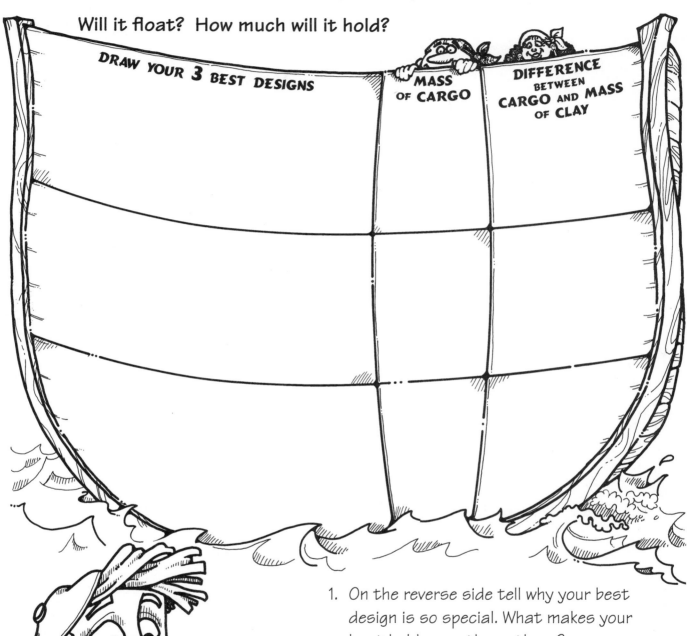

| DRAW YOUR 3 BEST DESIGNS | MASS OF CARGO | DIFFERENCE BETWEEN CARGO AND MASS OF CLAY |
|---|---|---|
| | | |
| | | |
| | | |

1. On the reverse side tell why your best design is so special. What makes your boat hold more than others?

2. Why does your boat float?

## Connecting Learning

1. For the designs that floated, which had more mass: the boat or the cargo?

2. For those that sank, which had more mass?

3. Is there a critical weight factor? Explain.

4. What other factors may have determined whether the boats would float?

5. What are you wondering now?

# How Much CARGO Will it Hold?

## Topic
Volume, mass, and density

## Key Question
If you owned an ocean liner, would you charge for hauling cargo by its mass, volume, or density? Why?

## Learning Goal
Students will compare mass, volume, and density measures to determine which should be used for making fair charges for hauling cargo.

## Guiding Documents
*Project 2061 Benchmarks*
- *Mathematics is helpful in almost every kind of human endeavor—from laying bricks to prescribing medicine or drawing a face. In particular, mathematics has contributed to progress in science and technology for thousands of years and still continues to do so.*
- *The expression a/b can mean different things: a parts of size 1/b each, a divided by b, or a compared to b.*

*NCTM Standards 2000\**
- *Develop strategies to determine the surface area and volume of selected prisms, pyramids, and cylinders*
- *Work flexibly with fractions, decimals, and percents to solve problems*
- *Collect data using observations, surveys, and experiments*
- *Solve simple problems involving rates and derived measurements for such attributes as velocity and density*

## Math
Rational numbers
  decimals
Using formulae
  area of circle
  volume of cylinder
  density
Calculating averages

## Science
Physical science
  density

## Integrated Processes
Observing
Collecting and recording data
Interpreting data
Generalizing
Applying

## Materials
Variety of cans (at least three different)
Metric rulers with mm markings
Balance
Masses
Salt, flour, rice, oats, beans

## Background Information
Students must understand that the amount of cargo a ship can carry is related to the volume of cargo area available and the mass that the ship can float.

From calculating the densities of the cargoes (salt, flour, rice, oats, and beans), the students should realize that if the same rate were applied to all cargo, the ship would not fairly earn all that it could. If the rate was only to be based on mass, light materials, especially those with a density less than 1, would fill the ship causing it to earn less than its maximum capacity. If the rate was only to be based on volume, heavy materials would weigh down the ship without using all the volume capacity, so the ship would earn less than it should. Through a class discussion, students should be able to generate a price structure based on the density of the cargo to be shipped.

Density is the ratio of mass to volume. It is calculated by dividing mass in grams by volume in milliliters or cubic centimeters.

The area of a circle is calculated by pi ($\pi$) times the radius to the second power ($A = \pi \times r^2$).

Volume of a cylinder is calculated by multiplying the area of the base (circle) times the height of the cylinder.

## Management

1. Be sure that cans are numbered so that students can refer back to the same can at different parts of the activity.
2. Time allowed should be one to two class periods.
3. Have students work in groups of four.
4. Have the cargo materials (salt, flour, rice, oats, and beans) in containers at one table.

5. Have the balances at another table.

## Procedure

1. Have students measure the diameter and height of a can and record in the data table.
2. Direct them to calculate the volume of the can.
3. Allow time for students to repeat steps 1 and 2 with two different cans.
4. Have students find the mass of the empty can.
5. Direct them to fill the can with one of the cargo materials.
6. Have them find the mass of the full container. Direct them to find the difference in the empty can's mass and the full can's mass to find the mass of the cargo.
7. Have students calculate the density of the cargo.
8. Allow students time to repeat steps 4-7 with each cargo type.
9. Use discussion time to determine the rates for each cargo.

## Connecting Learning

1. In which instances would the owner of the shipping company make more money charging by the mass of the cargo? ...volume of the cargo?
2. Why would the shipping company be better off charging by the cargo's density?
3. Using your density calculations, for which cargo would you charge the most? ...the least? Explain.
4. What are you wondering now?

## Extension

The consideration of liquid cargoes and/or salt water voyages which increase the weight capacity of the ship several percent gives the opportunity to extend the discussion or even recalculate the price chart.

\* Reprinted with permission from *Principles and Standards for School Mathematics,* 2000 by the National Council of Teachers of Mathematics. All rights reserved.

# How Much CARGO Will it Hold?

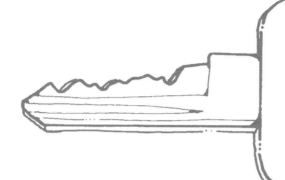

## Key Question

If you owned an ocean liner, would you charge for hauling cargo by its mass, volume, or density? Why?

## Learning Goal

### Students will:

compare mass, volume, and density measures to determine which should be used for making fair charges for hauling cargo.

# How Much CARGO Will it Hold?

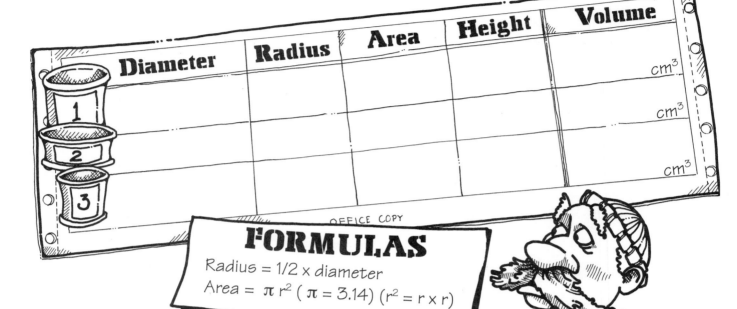

| | Diameter | Radius | Area | Height | Volume |
|---|---|---|---|---|---|
| 1 | | | | | cm³ |
| 2 | | | | | cm³ |
| 3 | | | | | cm³ |

OFFICE COPY

## FORMULAS

Radius = 1/2 x diameter

Area = $\pi r^2$ ( $\pi$ = 3.14) ($r^2$ = r x r)

1. Find the mass of the empty can. Fill it with cargo and find the new mass. Find the difference in the two masses. This difference is the mass of the cargo.
2. Divide the mass by the volume of cargo. This is the cargo's density.

| Cargo | Mass of Can Empty | Mass of Can Full | Mass of Cargo | Cargo Volume | Cargo Density |
|---|---|---|---|---|---|
| Salt | | | | | |
| Flour | | | | | |
| Rice | | | | | |
| Oats | | | | | |
| Beans | | | | | |

# How Much CARGO Will it Hold?

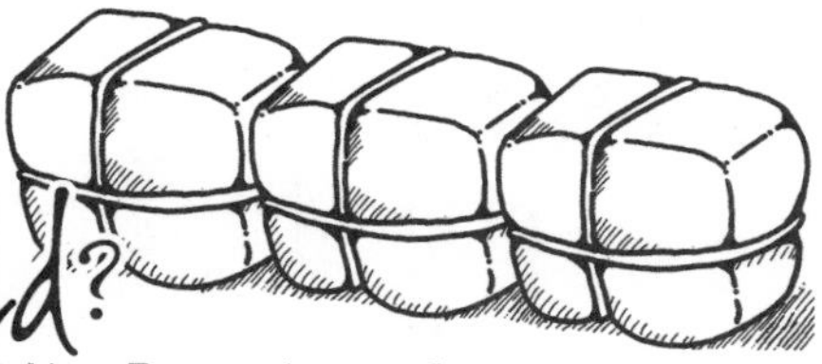

These rates are for
one cubic _____ .

## CARGO RATES

| CARGO | RATE |
|-------|------|
| Salt | $ |
| Flour | $ |
| Rice | $ |
| Oats | $ |
| Beans | $ |

My ship, The _____,
_____
holds _____
cubic _____ .

OFFICE COPY

129

# How Much CARGO Will it Hold?

## Connecting Learning

1. In which instances would the owner of the shipping company make more money charging by the mass of the cargo? …volume of the cargo?

2. Why would the shipping company be better off charging by the cargo's density?

3. Using your density calculations, for which cargo would you charge the most? …the least? Explain.

4. What are you wondering now?

# Topic
Problem solving

# Key Question
What is the largest amount of cargo you can float using one piece of aluminum foil?

# Learning Goal
Students will design an aluminum foil boat to hold the maximum amount of cargo.

# Guiding Documents
*Project 2061 Benchmark*
• *Calculate the circumferences and areas of rectangles, triangles, and circles, and the volumes of rectangular solids.*

*NCTM Standards 2000\**
• *Work flexibly with fractions, decimals, and percents to solve problems*
• *Collect data using observations, surveys, and experiments*

# Math
Multiplying decimals
Computing area and volume

# Science
Physical science
    buoyancy

# Integrated Processes
Observing
Collecting and recording data
Interpreting data
Generalizing
Applying

# Materials
Aluminum foil, heavy duty
Tape
Scissors
Metric rulers
Materials for cargo of known mass (e. g., centicubes)
Tubs for floating boats
Water
Paper towels

# Background Information
The volume of the ship is calculated by multiplying the area of the bottom of the ship times the draft (depth of the boat below the water line). Archimede's principle of buoyancy tells us that the weight that can be floated is equal to the weight of the water displaced by the vessel.

# Management
1. Time needed: one class period
2. Groups of two are best.
3. Each boat will be made from a 10 cm x 10 cm piece of aluminum foil. Provide enough foil that students can experiment with several designs.
4. Have plenty of towels handy. If the cargo sinks, it will need to be dried before being reused.
5. The inspector can be the teacher or a student from another group.

# Procedure
1. Using paper, have the students experiment with designs and calculations to find the best design for their boat.
2. Direct them to carefully construct their boat so that it is water tight and stable in the water.
3. Have them calculate the amount of cargo that their ships will carry. Tell them not to forget that the calculation of volume gives the total amount the ship will float; the mass of the ship itself must be included in that amount.
4. Have students fill out the vessel license application and present the application and vessel to the inspector prior to the launching.
5. The inspector will fill out the *For Office Use Only* portion of the activity page.

## Connecting Learning

1. What was the best design?
2. How does this design apply to freighters and barges?
3. What happens to the volume of the boat when you reduce the area of its base? [You reduce the total volume.]
4. What problem(s) might result from having a boat with low sides? [Waves might easily wash over it.]
5. What are wondering now?

## Extensions

1. Once the barge is constructed, the students can also calculate how much the barge could carry in salt water rather than in fresh water.
2. Straws can be used for structural reinforcement of the barge. If this reinforcement is allowed, a building regulation should be enforced that all supplementary material must be inside the foil hull; otherwise, straws can be sealed and used as flotation pontoons.
3. To create an ocean, mix a salt water solution by dissolving 200 mL of salt per 1 liter of water. Have the students calculate the density of that solution. Density is calculated by finding the mass of a sample of salt water and dividing the mass by the volume of the solution. The density is in g/mL, so the flotation capacity of the ship would be obtained by multiplying the ship's volume times the density of the water.

\* Reprinted with permission from *Principles and Standards for School Mathematics,* 2000 by the National Council of Teachers of Mathematics. All rights reserved.

# Ship Shape

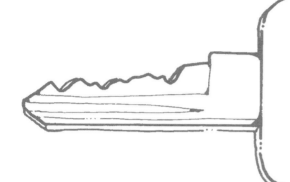

## Key Question

What is the largest amount of cargo you can float using one piece of aluminum foil?

## Learning Goal

### Students will:

design an aluminum foil boat to hold the maximum amount of cargo.

# Ship Shape

## Ship Use Permit Application

Ship's Name:

Ship's Dimensions:

Length:_____ cm

Width: _____ cm

Draft:_____ cm

Calculated Deck Area: _____cm$^2$

Volume of Cargo Area: _____ cm$^3$

Mass of Ship When Empty: _____ grams

Calculated Maximum Cargo Mass: _____ grams

Number of Cargo Items to be allowed:_____

Description of Cargo Items:

- - - - - - - - - - - - - - - - - - - - - - - - - - - - - - - - - - - - - - - - - - - - - - - - - - - - - - - - - -

### •For Office Use Only•

Test Data:

Tested Mass Carried _____ grams

Leakage: _____

Stability When Loaded:_____

_____

Percentage of Calculated Maximum Cargo Mass Allowed: _____ %

_____

Inspector's Signature                                    Date

# Ship Shape

## Connecting Learning

1. What was the best design?

2. How does this design apply to freighters and barges?

3. What happens to the volume of the boat when you reduce the area of its base?

4. What problem(s) might result from have a boat with low sides?

5. What are wondering now?

# Where Do You Draw the Line?

**PART 1**

## Topic
Volume and density

## Key Questions
1. How can you determine the level of the waterline on the hull of a loaded cargo ship?
2. Given a desired waterline, can you calculate how much cargo must be loaded into the cargo ship?

## Learning Goals
*Students will:*
1. calculate the location of the waterline when a loaded cargo ship is launched into fresh water, and
2. determine how much cargo the ship can carry so that the waterline is at a pre-determined level.

## Guiding Documents
*Project 2061 Benchmarks*
- *Organize information in simple tables and graphs and identify relationships they reveal.*
- *Read simple tables and graphs produced by others and describe in words what they show.*

*NCTM Standards 2000\**
- *Work flexibly with fractions, decimals, and percents to solve problems*
- *Collect data using observations, surveys, and experiments*
- *Develop strategies to determine the surface area and volume of selected prisms, pyramids, and cylinders*

## Integrated Processes
Observing
Predicting
Gathering and recording data
Interpreting data
Designing a procedure
Applying

## Math
Measuring
  height
  diameter
  mass
Calculating
  density
  area of a circle
  volume of a cylinder
  a proportional measure
Using formulae

## Science
Physical science
  density

## Materials
*For the class: (per group)*
  tin can (see *Management*)
  container of water large enough to float tin can
  fine sand or salt for cargo
  rubber band

## Background Information
In this activity, students will calculate the volume and density of the cargo ship. By comparing its density to the density of water, they will use a rubber band to mark the flotation line on the ship and test their predictions by placing the loaded cargo ship in the water.

Density is the measure of mass per unit volume. In this case: $D = m/v$ (g/mL). Students will need to determine the volume of the cylinder by multiplying the area of its base by its height. The area of the circular base is determine by multiplying $r^2$ (where r = radius of the base) and pi = 3.14.

Water at 4° C has the approximate density of 1.0 (1 g/mL). By comparing the density of the tin-can cargo ship to the water it will be floated in, you can figure how much of the can will be in the water. The ratio of the densities will give you what portion of the can will be under water.

The ratio $\dfrac{\text{density of ship}}{\text{density of water}} = \dfrac{?}{1.0} = x$

x = the amount of the ship that is under water.

To calculate the draft of the ship (the depth of the can below the waterline), each group will measure the height of the can, multiply that amount by x, then measure that distance from the bottom of the can and mark it with a rubber band.

## Management
1. This activity has two parts: The first is more structured in its format, while the second allows students to design a procedure to solve a situation.
2. Cans with flat bottoms are necessary for this activity. Small fruit juice cans work best. The volume of the cans will be determined using measurements taken on the exterior of the cans.
3. The cargo must be fine sand, salt, or other fine substances that will not shift positions and upset the ship.
4. This activity is best done in pairs or groups of four.
5. Students should have some prior experience with determining densities. Activities such as *Wat-ar Densities?*, *Floating Wood*, and *Afloat* are useful for establishing a conceptual framework.

## Procedure
*Part One*
1. Review the concept of density.
2. Ask *Key Question 1:* Can you determine the level of the waterline on the hull of the loaded tin-can cargo ship? Distribute tin-can ships and cargo.

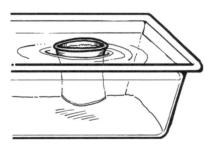

3. Have students fill the hold of the tin-can ship about 2/3 full of cargo.
4. Have them determine the mass, the volume, and the density of the combined ship and cargo. (The volume measure should be of the entire can).
5. By forming a ratio of the density of the ship to the density of the water (assumed to be 1.0), students should be able to predict the waterline on the hull when the ship is launched in their fresh water tank. Have them wrap a rubber band around the hull to mark the predicted waterline.
6. Have students test their predictions in a container of water. If predictions are not proximate, have students recalculate mass, volume, and density of their ships.

7. Repeat the process with two other types of cargo, one of which students may select.
8. Have students write about their accuracy in being able to predict the waterline and justify why they were or weren't able to accurately do so.

*Part Two*
1. Ask the second *Key Question:* Given a desired waterline, can you calculate how much cargo must be loaded into the tin-can cargo ship?
2. Inform students that they are to determine how much cargo is required to make the tin-can ship float at a waterline two centimeters from the top of the hull.
3. Because there is not a table to record data, students must construct one.
4. After making the determination of the amount of cargo necessary, have students launch their ships to test their predictions.

## Connecting Learning
1. Why did your ship float? [Its density was less than that of the water.]
2. Did the material you used for cargo make any difference in how the ship floated? Explain.
3. What would happen if you used water as your cargo? Would the ship float? Explain.
4. What information did you need to know in order to float your ship at a pre-determined waterline? How did you arrange your table?
5. Did any problems arise in this activity? If so, what were they? Can you devise a plan to overcome these problems?
6. What are you wondering now?

## Extensions
1. Have students investigate the history of Plimsoll lines used on ships.
2. Using different cargo materials, compare the load level of the hull for a pre-determined waterline.
3. Find a cargo that will completely fill the hull yet allow the ship to float at a pre-determined waterline.

* Reprinted with permission from *Principles and Standards for School Mathematics,* 2000 by the National Council of Teachers of Mathematics. All rights reserved.

# Where Do You Draw the Line?

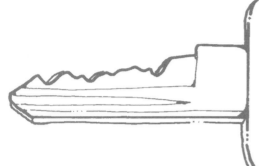

## Key Questions

1. How can you determine the level of the waterline on the hull of a loaded cargo ship?
2. Given a desired waterline, can you calculate how much cargo must be loaded into the cargo ship?

## Learning Goal

### Students will:

1. calculate the location of the waterline when a loaded cargo ship is launched into fresh water, and
2. determine how much cargo the ship can carry so that the waterline is at a pre-determined level.

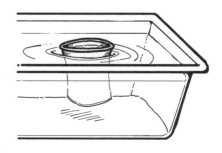

138

# Where Do You Draw the Line?

**PART 1**

Can you predict the level of the waterline on the hull of a loaded tin-can cargo ship?

1. Fill the hold of a tin-can ship about $^2/_3$ full of cargo consisting of fine sand, salt, birdseed, or similar substances.

2. Determine the density of the combined ship and cargo.

3. Calculate the level of the waterline on the hull when the ship is launched in fresh water. Wrap a rubber band around the hull to mark the predicted waterline.

4. Test your prediction by floating the tin-can ship.

5. Repeat with two other types of cargo.

Remember, $V = \pi r^2 h$ and $D = m/v$

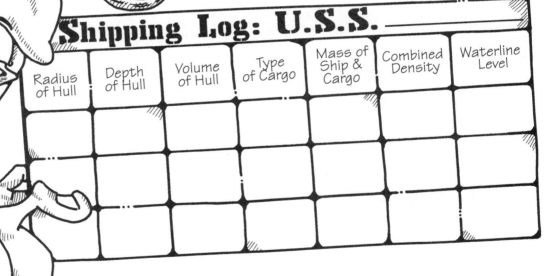

| Shipping Log: U.S.S. _____ | | | | | | |
|---|---|---|---|---|---|---|
| Radius of Hull | Depth of Hull | Volume of Hull | Type of Cargo | Mass of Ship & Cargo | Combined Density | Waterline Level |
| | | | | | | |
| | | | | | | |
| | | | | | | |

Could you have predicted the waterline more accurately? Explain.

# Where Do You Draw the Line?

Part 2

Can you calculate how much cargo must be loaded into the tin-can cargo ship so its waterline will be at predetermined level?

1. Wrap a rubber band around the hull about 2 cm from the top.
2. Design a table that contains information necessary for solving this problem.
3. Determine the amount of cargo needed to reach the desired waterline.
4. Add the proper amount of cargo.
5. To test, launch the ship.

Shipping Log: U.S.S. _____

Describe the results:

# Where Do You Draw the Line?

**PART 1**

## Connecting Learning

1. Why did your ship float?

2. Did the material you used for cargo make any difference in how the ship floated? Explain.

3. What would happen if you used water as your cargo? Would the ship float? Explain.

4. What information did you need to know in order to float your ship at a pre-determined waterline? How did you arrange your table?

5. Did any problems arise in this activity? If so, what were they? Can you devise a plan to overcome these problems?

6. What are you wondering now?

## Topic
Mass

## Key Question
What is the mass of each object?

## Learning Goal
Students will find the mass of various objects and compare that mass to the size (volume) and shape of an object.

## Guiding Documents
*Project 2061 Benchmarks*
- *Find what percentage one number is of another and figure any percentage of any number.*
- *Use, interpret, and compare numbers in several equivalent forms such as integers, fractions, decimals, and percents.*

*NCTM Standards 2000\**
- *Collect data using observations, surveys, and experiments*
- *Work flexibly with fractions, decimals, and percents to solve problems*
- *Solve simple problems involving rates and derived measurements for such attributes as velocity and density*

## Math
Estimating
Measuring
   mass
Whole number operations
Rational numbers
   percents

## Integrated Processes
Observing
Classifying
Collecting and organizing data
Interpreting data
Generalizing

## Materials
*An assortment of the following or similar objects:*
   pack of gum
   box of snack mix
   individual box of raisins
   box of animal crackers
   a caramel candy
   package of powdered gelatin
   box of facial tissues
   crayons in a box
   box of chalk
   pencil box
   box of staples
   box of paper clips
Balance
Masses

## Background Information
Before working with density, students need to be acquainted with the idea of mass. Mass is the measure of the quantity of matter contained in a body. In this activity, students should note that mass is a separate notion from the notions of size (volume) of an object. This investigation will also allow students more familiarity with the metric unit of grams.

## Management
1. This activity is appropriate for either large or small group instruction.
2. Since this can be done rather quickly, groups could be rotated through stations.
3. All masses on packages should be inked over or covered.
4. If more than ten objects are used, ordering becomes more difficult.

## Procedure

1. Without touching the objects, have students predict the order of objects by mass from lightest to heaviest.
2. Ask students to estimate the mass of each object and record.
3. Have them find the actual mass of each object, calculate the difference in their estimation and the actual, and figure the percentage of error (Difference/Actual x 100 = % of Error).

## Connecting Learning

1. How did you do when you predicted the ordering of the objects?
2. Could you have correctly predicted the order if you could have touched the objects? Explain.
3. What objects did you have the least difference between the actual and the estimated mass?
4. What object had the greatest difference? Why do you think this happened?
5. Did the object with the greatest volume also have the greatest mass?
6. What object, not represented here, can you think of that would have a very large volume but not have a very large mass?
7. What object, not represented here, can you think of that would have a very small volume but a large mass?
8. What do you think this investigation is trying to help you learn?
9. What are you wondering now?

## Extensions

1. Rank student estimates (+ and -) to show the range of estimates. Sum them for comparisons of class estimates of each object.
2. Find the average for percent error to rank student's estimates.

\* Reprinted with permission from *Principles and Standards for School Mathematics,* 2000 by the National Council of Teachers of Mathematics. All rights reserved.

# MASSIVE BOXES

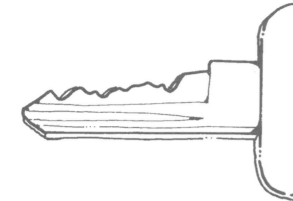

## Key Question

What is the mass of each object?

## Learning Goal

### Students will:

find the mass of various objects and compare that mass to the size (volume) and shape of an object.

# MASSIVE BOXES

**Predictions**

Which object has:
- the smallest volume?
- the least density?
- the largest volume?
- the greatest density?

**Compute the volume and density of the 10 objects.**

| Object | Mass (grams) | Volume (cm³) | Density (g/cm³) |
|---|---|---|---|
| 1. | | | |
| 2. | | | |
| 3. | | | |
| 4. | | | |
| 5. | | | |
| 6. | | | |
| 7. | | | |
| 8. | | | |
| 9. | | | |
| 10. | | | |

**Describe your findings.**

Were there any surprises?

# MASSIVE BOXES

### Connecting Learning

1. How did you do when you predicted the ordering of the objects?

2. Could you have correctly predicted the order if you could have touched the objects? Explain.

3. What objects did you have the least difference between the actual and the estimated mass?

4. What object had the greatest difference? Why do you think this happened?

5. Did the object with the greatest volume also have the greatest mass?

6. What object, not represented here, can you think of that would have a very large volume but not have a very large mass?

7. What object, not represented here, can you think of that would have a very small volume but a large mass?

8. What do you think this investigation is trying to help you learn?

9. What are you wondering now?

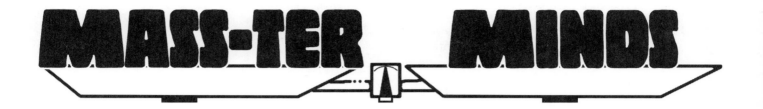

## Topic
Mass of spherically-shaped foods compared to their volume

## Key Question
Does size (volume) tell you heaviness (mass)?

## Learning Goal
Students will compare the volumes and masses of various objects.

## Guiding Documents
*Project 2061 Benchmark*
- *Organize information in simple tables and graphs and identify relationships they reveal.*

*NCTM Standards 2000\**
- *Collect data using observations, surveys, and experiments*
- *Work flexibly with fractions, decimals, and percents to solve problems*
- *Solve simple problems involving rates and derived measurements for such attributes as velocity and density*

## Math
Whole number operations
Averaging
   mean
Estimating
Measuring
   mass
Using rational numbers
   decimals
   percents

## Science
Physical science
   density

## Integrated Processes
Observing
Predicting
Classifying
Collecting and recording data
Interpreting data
Generalizing
Applying

## Materials
Jaw breakers, oranges, grapefruits, popcorn balls, tomatoes, grapes, cranberries, onions or any other spherical-shaped fruits, vegetables, or foods
Balances
Masses

## Background Information
Density is the measure of mass per unit volume. It is defined by the formula $D = m/v$. Students need to have an intuitive feeling for mass. This should be acquired through this investigation. They will discover that something is not heavier just because it is bigger, but that there is something else to consider. That something is density.

## Management
1. Allow approximately one class period for the basic investigation.
2. This can be done as a whole class activity. Have a small group of students lead the class. Let one hold up the object while the remainder of the class writes down their estimates. While this is going on, another student can be putting the diagram on the board. Two more students can be setting up the balances and finding the mass of each object as the leader finishes showing it. Percent error can be calculated during another period if desired.
3. To find the class average estimate, divide students into groups of four or five. Have them find their group's average and share it with the class. The class average can be calculated from the groups' data.

## Procedure

1. Using sight only, have the students write down their predictions of the objects from lightest mass to heaviest mass.
2. Have them record their estimates of the mass of each object in grams.
3. Direct the students to find the actual mass and the difference between the actual and the estimated values.
4. Have them compute the percent error for each object.
5. Direct them to find and record the class average estimate for each object and the percent of error.

## Connecting Learning

1. How did your predicted order compare with the actual order?
2. Which objects had greater masses than you thought? Which had less?
3. Do you think the largest one (volume) always has the greatest mass? Explain.
4. If these all had the same volume, would the order of mass change? How could we find out? [Find their densities.]
5. What are you wondering now?

## Extension

After using edible objects, have the students try this with balls made of different materials.

\* Reprinted with permission from *Principles and Standards for School Mathematics,* 2000 by the National Council of Teachers of Mathematics. All rights reserved.

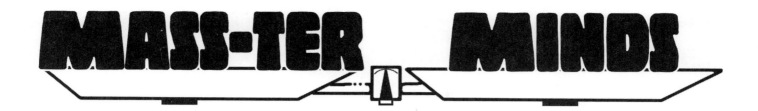
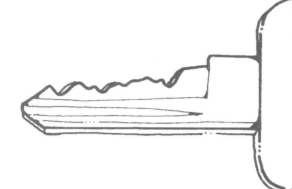

## Key Question

Does size (volume) tell you heaviness (mass)?

## Learning Goal

**Students will:**

compare the volumes and masses of various objects.

# MASS-TER MINDS

| Object | Mass Estimate | Actual Mass | Difference |
|---|---|---|---|
| | | | |
| | | | |
| | | | |
| | | | |
| | | | |
| | | | |
| | | | |

| Object | % Error of My Guess | Class Average Guess | % Error of Class Guess |
|---|---|---|---|
| | | | |
| | | | |
| | | | |
| | | | |
| | | | |
| | | | |

**TO FIND % ERROR:**

1. Find the difference between your estimate and the actual.
2. Divide this answer by the actual mass and round to the nearest hundredth.
3. Move the decimal point two places to the right and add the percent sign. You're done!

# Connecting Learning

1. How did your predicted order compare with the actual order?

2. Which objects had greater masses than you thought? Which had less?

3. Do you think the largest one (volume) always has the greatest mass? Explain.

4. If these all had the same volume, would the order of mass change? How can we find out?

5. What are you wondering now?

# One Way or Another

$$V = 4/3 \pi r^3$$

## Topic
Volume of spheres and their densities

## Key Questions
1. How can we find the volume of a sphere?
2. Do objects with large volumes have greater densities?

## Learning Goal
Students will find the volumes of spherical objects by using two different methods: the water displacement method and application of the formula for finding the volume of a sphere, $V = 4/3 \pi r^3$.

## Guiding Documents
*Project 2061 Benchmarks*
- *Organize information in simple tables and graphs and identify relationships they reveal.*
- *Use, interpret, and compare numbers in several equivalent forms such as integers, fractions, decimals, and percents.*
- *Find what percentage one number is of another and figure any percentage of any number.*

*NCTM Standards 2000\**
- *Collect data using observations, surveys, and experiments*
- *Work flexibly with fractions, decimals, and percents to solve problems*
- *Solve simple problems involving rates and derived measurements for such attributes as velocity and density*

## Math
Measuring
   mass
   volume
Estimating
   rounding
Using formulae
   density
   volume of sphere
Rational numbers
   fractions
   decimals
   percent

## Science
Physical science
   density

## Integrated Processes
Observing
Predicting
Collecting and recording data
Interpreting data
Generalizing
Applying

## Materials
Spherical objects such as high density balls, table tennis
   balls, jack balls, golf balls, marbles
Graduated cylinders
Water
Balances
Masses
Wooden blocks, optional (see *Procedure*)
Metric rulers

## Background Information
   An object submerged in water will displace a volume of water equal to its own volume. To find volume using the *water displacement method*, fill a graduated cylinder to a level great enough that the object can be totally submerged. Note the volume of the water. Submerge the object and note the water level. The difference in the two levels is the volume of the object. (NOTE: 1 cubic centimeter equals 1 milliliter of water at 4°C.) Some objects may have to be pushed under the water level to displace the water. Be careful not to displace more than necessary. The densities of the spherical objects used in this activity will not have any consequence on the amount of water displaced; only the objects' volumes will affect the amount of water displaced.

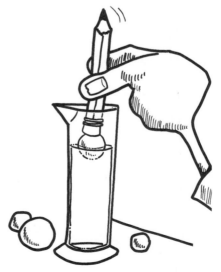

After finding the volumes of the spherical objects using the water displacement method, students are asked to apply the formula for finding the volume of a sphere, $V = 4/3 \pi r^3$. Comparisons between the answers determined by both methods will be made.

## Management
1. This can be done as a class activity or in small groups.
2. Suggested class time is about 45 minutes.
3. Make certain that the graduated cylinders are large enough in circumference to accommodate the spherical objects.
4. Gather the spherical objects before beginning this activity.
5. Each group will need five objects. To minimize the number of objects needed, groups may share the objects. The sharing of objects allows groups to compare their findings.

## Procedure
1. Have the students find and record the mass of each of their five spherical objects.
2. To find the volume of the spheres by water displacement, direct them to follow the directions on their activity sheet. Have them record the volume measures in milliliters.
3. Instruct students to find the densities of their objects using the masses they recorded and the volumes determined by water displacement. The units of density in this section will be g/mL.
4. Tell the students that they will be finding the volumes of the same objects this time using a formula.
5. Have them determine the volumes using the formula $V = 4/3 \pi r^3$. An easy way to obtain the diameter of a sphere is to place it between two blocks of wood and then measure the distance between the blocks. Half of that will be the radius.

6. Direct them to find the density by dividing the mass by the volume they just calculated. This time the densities will be recorded using the units g/cm³. If necessary, remind students that 1 mL of water (4°C) is equal to 1 cubic centimeter.

7. Allow time for students to compare their *Volume* results and decide which method they think is most accurate.
8. Have them determine their percentage of error by finding the difference in their volume measures for each object, dividing the difference by what they determined to be their most accurate measure, multiplying by 100 and adding a percent sign.
9. Allow the students to analyze their results until they conclude that the large spheres displace more water but are not necessarily heavier.

## Connecting Learning
1. Which object displaced the most water?
2. Which is the heaviest object?
3. Is the object that displaced the most water also the heaviest? Explain.
4. Which is the largest object?
5. Is the largest object always the one that displaces the most water? Explain.
6. What causes water displacement (size, heaviness, color)?
7. Can we use this method to find the volume of every spherical object? Explain.
8. Which method for finding volume did you determine to be most accurate? Why?
9. What are you wondering now?

## Extension
Obtain a vernier caliper and let the students measure the diameter of the object and then check the accuracy of the previous results.

* Reprinted with permission from *Principles and Standards for School Mathematics,* 2000 by the National Council of Teachers of Mathematics. All rights reserved.

# One Way or Another

$V = 4/3 \pi r^3$

## Key Questions

1. How can we find the volume of a sphere?
2. Do objects with large volumes have greater densities?

## Learning Goal

### Students will:

find the volumes of spherical objects by using two different methods: the water displacement method and application of the formula for finding the volume of a sphere,
$V = 4/3 \pi r^3$.

## VOLUME BY DISPLACEMENT

1. Find and record the mass of the objects.
2. Fill a graduated cylinder to a certain level and record this level as Level 1.
3. Place object in water and read new water level. Record the new level as Level 2.
4. Subtract to find the difference in the two levels. This difference is equal to the volume of the object.
5. Divide the mass by the volume to obtain density.

| Object | Mass (g) | Level 1 (mL) | Level 2 (mL) | Volume (mL) | Density (g/mL) |
|---|---|---|---|---|---|
| | | | | | |
| | | | | | |
| | | | | | |
| | | | | | |
| | | | | | |

# One Way or Another

$V = 4/3 \pi r^3$

## VOLUME WITH FORMULA

1. Record the mass of each object.
2. Measure the diameter (cm) of each object and divide by 2 to determine the radius.
3. Multiply radius x radius x radius. Record your answer in $cm^3$.
4. To find the volume of the sphere, use this formula: $V = {}^4/_3 \pi r^3$. (Let $\pi = {}^{22}/_7$ or 3.14.) Record.
5. Divide the mass by the volume to obtain density.

### TO FIND PERCENT OF ERROR

1. Find the difference in density from Volume By Displacement and Volume with Formula.
2. Divide the difference by the density you think is more accurate.
3. Multiply your answer by 100 and add a percent sign.

| Object | Mass (g) | Diameter (cm) | Radius (cm) | Radius³ (cm³) | Volume (cm³) | Density g/cm³ | % Error |
|--------|----------|---------------|-------------|---------------|--------------|---------------|---------|
|        |          |               |             |               |              |               |         |
|        |          |               |             |               |              |               |         |
|        |          |               |             |               |              |               |         |
|        |          |               |             |               |              |               |         |
|        |          |               |             |               |              |               |         |

Which method (water displacement or use of a formula) do you think was most accurate for determining the volume of the spheres? Explain.

# One Way or Another

$$V = 4/3 \, \pi \, r^3$$

## Connecting Learning

1. Which object displaced the most water?

2. Which is the heaviest object?

3. Is the object that displaced the most water also the heaviest? Explain.

4. Which is the largest object?

5. Is the largest object always the one that displaces the most water? Explain.

6. What causes water displacement? ...size? ...heaviness? ...color?

7. Can we use this method to find the volume of every spherical object? Explain?

8. Which method for finding volume did you determine to be most accurate? Why?

9. What are you wondering now?

# Will it Float?

## Topic
Density

## Key Question
Will an orange float or sink in water?

## Learning Goal
Students will determine the approximate density of an orange and then predict whether it will float or sink in water.

## Guiding Documents
*Project 2061 Benchmarks*
- *Find what percentage one number is of another and figure any percentage of any number.*
- *Decide what degree of precision is adequate and round off the result of calculator operations to enough significant figures to reasonably reflect those of the inputs.*

*NCTM Standards 2000\**
- *Collect data using observations, surveys, and experiments*
- *Work flexibly with fractions, decimals, and percents to solve problems*
- *Select and apply techniques and tools to accurately find length, area, volume, and angle measures to appropriate levels of precision*
- *Solve simple problems involving rates and derived measurements for such attributes as velocity and density*

## Math
Measuring
    volume
    circumference
Using formulae
    volume
    density
Rational numbers

## Science
Physical science
    density

## Integrated Processes
Observing
Predicting
Collecting and recording data
Interpreting data
Applying

## Materials
*Per group:*
    large orange
    measuring tape
    metric ruler
    balance
    masses
    liter box or large graduated cylinder
    paper towels

## Background Information
Density is the ratio of mass to volume: $D=m/v$. Water at 4°C has a density of one. Objects that have a density less than one will float. Those with a density greater than one will sink, unless they are held up by the surface tension of water or their shape allows them to displace an amount of water equal to their weight.

The whole orange usually has a density less than one and will float. When peeled, the orange sections will sink (unless they have been damaged by freezing) and the peel will float. The orange sections, while they consist primarily of water, also have a significant amount of sugar and other substances that increase their density. The peeling has a fibrous structure with a density considerably less than one causing it to float. In a sense, the peeling acts as a "life jacket" for the orange.

Students can test the peel and the section at the conclusion of the activity to determine which floats and which sinks. If the sections are left together, make sure there are no air pockets trapped in the core.

## Management

1. This investigation requires little preparation of materials.
2. Decide whether to have students follow the guide on the student page or develop their own procedures. If the unstructured approach is to be used, introduce the activity by asking students, "How could we determine whether an orange would float without first placing it in water?" Allow time for the formulation of procedures if this unstructured approach is used.

## Procedure

1. Divide the students into groups.
2. Determine whether the structured or unstructured approach is to be used.
3. Ask students to design their procedures if the unstructured approach is used.
4. Provide each group with the required materials and proceed with the investigation.
5. Discuss why the use of a ruler inserted into the liter box permits a more accurate determination of the height of the water. [It permits reading in ten mL intervals rather than one hundred mL intervals.]

## Connecting Learning

1. Does the unpeeled orange float or sink in water?
2. Was your prediction correct?
3. Does the orange peel float or sink? Explain why you think it does this.
4. Do the orange sections float or sink? Explain why you think they do this.
5. Did all the sections behave the same way?
6. How did the results from the two methods of determining the volume compare? Which do you think is more accurate? Why?
7. What are you wondering now?

## Extensions

1. Have the students calculate the densities of other fruits and test whether they float or sink.
2. Find different varieties of oranges and compare their densities (navel and Valencia, for example).

*   Reprinted with permission from *Principles and Standards for School Mathematics,* 2000 by the National Council of Teachers of Mathematics. All rights reserved.

## Key Question

Will an orange float or sink in water?

## Learning Goal

determine the approximate density of an orange and then predict whether it will float or sink in water.

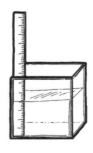

# Will it Float?

**Will an orange sink or float in water?** Your prediction: _____

An orange is only approximately round. However, it is possible to use the formula for the volume of a sphere if we find an average radius.

To find an average radius, take one measure of the orange's "polar circumference" and one of its "equatorial circumference" and find the average. From that, calculate the average radius.

| | | |
|---|---|---|
| Polar circumference | _____ cm |
| Equatorial circumference | _____ cm |
| Average circumference | _____ cm |
| Average radius ($\frac{c}{2\pi}$) | _____ cm |
| Volume ($V = \frac{4}{3}\pi r^3$) | _____ cm |
| Mass of Orange | _____ grams |
| Density of Orange ($D = \frac{m}{v}$) grams per mL | |

Test of your prediction:
Place the orange in water. Is your prediction right?

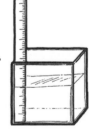

Measure the displacement of water by an orange. Place a ruler into the liter box to provide a more accurate scale for reading volume.

What is the different between the calculated volume of the orange and the volume as determined by displacement? _____mL

By what percent does the calculated volume differ from the displacement volume measurement? _____%

# Will it Float?

## Connecting Learning

1. Does the unpeeled orange float or sink in water?

2. Was your prediction correct?

3. Does the orange peel float or sink? Explain why you think it does this.

4. Do the orange sections float or sink? Explain why you think they do this.

5. Did all the sections behave the same way?

6. How did the results from the two methods of determining the volume compare? Which do you think is more accurate? Why?

7. What are you wondering now?

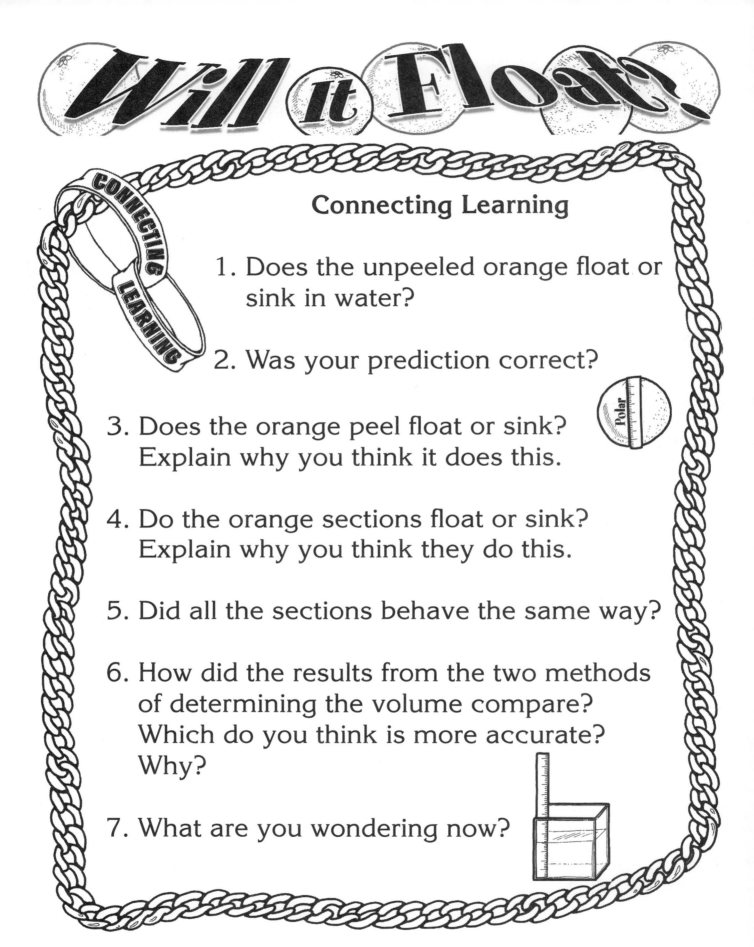

# PLAY BALL!

## Topic
Density

## Key Questions
1. If the sports balls used in this investigation are arranged in order from the lowest to highest density, what will that order be?
2. What is the density of each sports ball?

## Learning Goal
Students will determine the densities of six different sports balls.

## Guiding Documents
*Project 2061 Benchmarks*
- *Organize information in simple tables and graphs and identify relationships they reveal.*
- *Calculate the circumferences and areas of rectangles, triangles, and circles, and the volumes of rectangular solids.*

*NCTM Standards 2000\**
- *Collect data using observations, surveys, and experiments*
- *Work flexibly with fractions, decimals, and percents to solve problems*
- *Select and apply techniques and tools to accurately find length, area, volume, and angle measures to appropriate levels of precision*
- *Solve simple problems involving rates and derived measurements for such attributes as velocity and density*

## Math
Estimating
Measuring
  circumference
  mass
Using formulae
  $C = d\pi$
  $V = 4/3 \, \pi \, r^3$
  $D = m/v$
Graphing

## Science
Physical science
  density

## Integrated Processes
Observing
Predicting
Collecting and recording data
Interpreting data

## Materials
*Each group will examine six balls:*
  tennis ball
  basketball
  baseball
  softball
  volleyball
  soccer ball
  rubber ball
  table tennis ball
  golf ball
  croquet ball
  racket ball
Metric tape measures
Balance
Masses

## Background Information
The formula for finding the volume of a sphere is $V = 4/3 \, \pi \, r^3$. To find the volume of a sphere, we need to know the radius. By measuring the circumference of the sphere, we can determine the sphere's diameter by dividing the circumference (C) by $\pi$ : $d = C/\pi$. The radius is then found by dividing the diameter by 2 or it can be found directly by the formula $r = C/2\pi$.

## Management
1. Have students collect a variety of sports balls.
2. Each group, or team, will need six. Interest is increased if the groups have as many different kinds of balls as possible.

## Procedure
1. Have each team select six sports balls for their investigation.
2. Working as teams, have each group rank order the sports balls from lowest to highest density by estimating. Also, have each member of the group estimate the density of each ball.
3. Direct the teams to find the circumference and mass of each ball, and record the information in the table.
4. Working individually and then comparing results, have students find the volume of each ball.
5. Direct them to work individually to find the density of each ball. Have them compare their results.
6. Allow the teams time to compare their measured and computed results with their estimates and describe what they found.

## Connecting Learning
1. Were you able to rank order the sports balls? Why do you think this is difficult to do?
2. Which ball was the densest?
3. Do you think sports equipment manufacturers are concerned about the density of the balls? Explain.
4. What are you wondering now?

## Extensions
1. Have students determine the radius of the balls by placing a ball between two blocks of wood and measuring the distance between the blocks to obtain the diameter of the ball. They should then divide the diameter by two to determine the radius. Have them compare the result with the radius that was determined in the activity using $d = C/2\pi$.
2. Have students write to different sports equipment manufacturers to investigate the role the densities of various balls.

\* Reprinted with permission from *Principles and Standards for School Mathematics,* 2000 by the National Council of Teachers of Mathematics. All rights reserved.

# PLAY BALL!

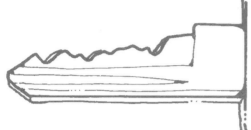

## Key Questions

1. If the sports balls used in this investigation are arranged in order from the lowest to highest density, what will that order be?
2. What is the density of each sports ball?

## Learning Goal

**Students will:**

determine the densities of six different sports balls.

# PLAY BALL!

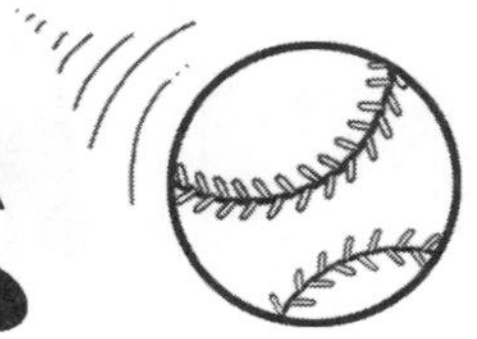

Members of the group are:

| Sports Balls We Used | Our Estimates of Densities | Actual Densities | Our Group's Rankings* | Actual Rankings |
|---|---|---|---|---|
| | | | | |
| | | | | |
| | | | | |
| | | | | |

*LOWEST IS 1 HIGHEST IS 6

These are our results:

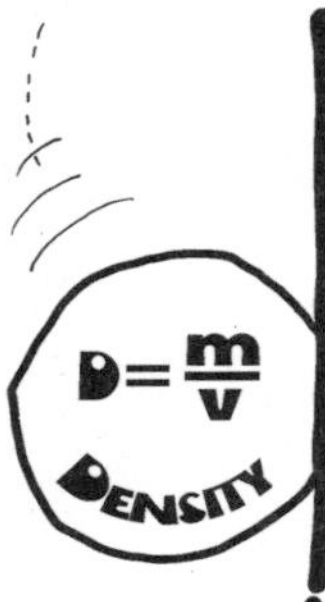

| Sport Ball | Mass (g) | Circumference (cm) | Diameter (cm) | Radius (cm) | Volume (cm$^3$) | Density |
|---|---|---|---|---|---|---|
| | | | | | | |
| | | | | | | |
| | | | | | | |
| | | | | | | |
| | | | | | | |

$D = \dfrac{m}{V}$
DENSITY

$d = \dfrac{c}{\pi}$
DIAMETER

$c = d\pi$
CIRCUMFERENCE

$r = \dfrac{d}{2}$
RADIUS

$V = \dfrac{4}{3}\pi r^3$
VOLUME

Here is what we discovered:

# PLAY BALL!

## Connecting Learning

1. Were you able to rank order the sports balls? Why do you think this is difficult to do?

2. Which ball was the densest?

3. Do you think sports equipment manufacturers are concerned about the density of the balls? Explain.

4. What are you wondering now?

$$d = \frac{c}{\pi}$$

DIAMETER

$$r = \frac{d}{2}$$

RADIUS

**Topic**
Density

**Key Questions**
1. Do plastic objects sink or float?
2. What is the density of plastic objects?

**Learning Goal**
Students will find and graph the densities of four plastic objects.

**Guiding Documents**
*Project 2061 Benchmarks*
- *Decide what degree of precision is adequate and round off the result of calculator operations to enough significant figures to reasonably reflect those of the inputs.*
- *Organize information in simple tables and graphs and identify relationships they reveal.*

*NCTM Standards 2000\**
- *Collect data using observations, surveys, and experiments*
- *Solve simple problems involving rates and derived measurements for such attributes as velocity and density*

**Math**
Measurement
    mass
    water displacement
    volume
Using formulae
    density
Graphing

**Science**
Physical science
    density

**Integrated Processes**
Designing an investigation (*Page 1*, only)
Observing
Predicting
Comparing and contrasting

Collecting and recording data
Interpreting data
Communicating

**Materials**
Paper towels
*For each group:*
    500 mL graduated cylinder
    balance
    masses
    50 AIMS Astronaut Counters
    50 Teddy Bear Counters
    6-8 each, 10g and 20g GramStackers
    130 AIMS Friendly Bears
    (Other **uniform** plastic objects may be substituted for any of the above.)

**Background Information**
This investigation may be done in either of two ways: completely open-ended except for the choice of objects or more structured. Use the first student page for the open-ended alternative and the remaining student pages for the structured alternative.

In this activity students will use water displacement to determine the volume of the irregularly-shaped plastics. The volume of the plastic pieces is equal to the volume of water displaced. The volume of water displaced is determined by reading the change in the water level.

This activity provides abundant practice in reading a graduated cylinder, finding the mass of objects, computing density and broken-line graphing. To complete the graph, have students draw the *one-line* which is the graph of the density of water. It makes a 45 degree angle with the horizontal axis. It is of interest to note that any broken line above the one-line is the graph of an object that sinks and any broken line below the one-line is the graph of an object that floats!

## Management

1. Students should be instructed to take great care in reading the water levels in the graduated cylinder. The cylinder should be set on a flat table top for each reading.
2. To reduce the quantity of plastic objects required, each group may work with one type of object at a time and then exchange. Objects should be toweled dry after each use.
3. Students may wish to determine the average mass of each type of object by first finding the total mass of a large number and then dividing to determine the average mass.
4. Inform students that splashing is minimized when adding the plastic objects by tilting the cylinder and letting the objects slide into the water.

## Procedure

1. Ask students to study the types of plastic objects and predict the order of density from least to greatest.
2. Have students fill the graduated cylinders to the 300 mL level at the outset of each part of the investigation. An eyedropper or syringe is convenient for making the final adjustment.
3. After finding all the data and entering it into the table, ask students to determine the density for each type of plastic and graph it with a broken-line graph. In this investigation, the unit for density is grams per milliliter (g/mL).

## Connecting Learning

1. How did your predicted order of density compare with the order obtained through this investigation?
2. What techniques were the most useful in obtaining accurate results for both the mass and the volume?
3. What does it mean when a broken-line density graph lies above or below the one-line?
4. What are you wondering now?

*   Reprinted with permission from *Principles and Standards for School Mathematics,* 2000 by the National Council of Teachers of Mathematics. All rights reserved.

# SINK or SWIM

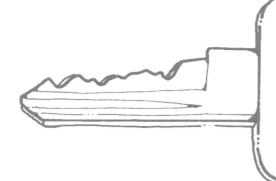

## Key Questions

1. Do plastic objects sink or float?
2. What is the density of plastic objects?

## Learning Goal

### Students will:

find and graph the densities of four plastic objects.

# SINK or SWIM

**Which of the following plastic objects do you think will float?**

**Using your best judgement, arrange them in order of density.**

| Object | Float | Sink |
|---|---|---|
| AIMS Astronaut Counters | _____ | _____ |
| Teddy Bear Counters | _____ | _____ |
| AIMS Gramstackers | _____ | _____ |
| AIMS Friendly Bears | _____ | _____ |

Least _____

_____

_____

Greatest _____

1. Use a 500 mL graduated cylinder filled with water to the 300 milliliter level.

2. Find the missing information.

3. If any of the objects float, use a flat-bottomed can or bottle to submerge the objects without letting it touch the water.

### ASTRONAUT COUNTERS

| Water Level | Displacement | Number of Counters | Total Mass of Counters | Density of Counters |
|---|---|---|---|---|
| 300 mL | | | | |
| 350 mL | | | | |
| 400 mL | | | | |
| 450 mL | | | | |
| 500 mL | | | | |

# SINK or SWIM

## TEDDY BEAR COUNTERS

| Number of Counters | Mass of Counters | Water Level | Net Displacement | Density of Counters |
|---|---|---|---|---|
| 0 | 0 | 300 mL | 0 | |
| 10 | | | | |
| 20 | | | | |
| 30 | | | | |
| 40 | | | | |

Use 10g and 20g masses.

## GRAMSTACKERS

| Number of Grams | Water Level | Net Displacement | Density of GramStackers |
|---|---|---|---|
| 0 | 300 mL | 0 | |
| 30 | | | |
| 60 | | | |
| 90 | | | |
| 120 | | | |
| 150 | | | |
| 180 | | | |

## FRIENDLY BEARS

| Water Level | Net Displacement | Number of Bears | Mass of Bears | Density of Bears |
|---|---|---|---|---|
| 300 mL | 0 | 0 | | |
| 350 mL | | | | |
| 400 mL | | | | |
| 450 mL | | | | |
| 500 mL | | | | |

# Sink or Swim

Rank order the objects from least to greatest density and compare the order with your prediction.

Least _____

_____

_____

Greatest _____

Construct a broken-line graph for each of the objects.
Graph the density of water for comparison.

# Sink or Swim

## Connecting Learning

1. How did your predicted order of density compare with the order obtained through this investigation?

2. What techniques were the most useful in obtaining accurate results for both the mass and the volume?

3. What does it mean when a broken-line density graph lies above or below the one-line?

4. What are you wondering now?

# What's in a BB?

## Topic
Density

## Key Question
Of what substance are BBs composed?

## Learning Goal
Students will find the density of BBs and match that against the density of metals to determine the substance of which they are made.

## Guiding Documents
*Project 2061 Benchmarks*
- *What people expect to observe often affects what they actually do observe. Strong beliefs about what should happen in particular circumstances can prevent them from detecting other results. Scientists know about this danger to objectivity and take steps to try and avoid it when designing investigations and examining data. One safeguard is to have different investigators conduct independent studies of the same questions.*
- *Mathematics is helpful in almost every kind of human endeavor—from laying bricks to prescribing medicine or drawing a face. In particular, mathematics has contributed to progress in science and technology for thousands of years and still continues to do so.*
- *Buttress their statements with facts found in books, articles, and databases, and identify the sources used and expect others to do the same.*

*NCTM Standards 2000\**
- *Collect data using observations, surveys, and experiments*
- *Solve simple problems involving rates and derived measurements for such attributes as velocity and density*

## Math
Measuring
    mass
    volume
Using formulae
    density

## Science
Physical science
    density

## Integrated Processes
Designing an investigation (*Page One* only)
Observing
Collecting and recording data
Interpreting data
Comparing and contrasting
Inferring
Applying

## Materials
*For each group:*
    balance
    masses
    500 mL graduated cylinder
    BBs sufficient to fill 150 mL graduated cylinder
    paper towels for drying BBs

## Background Information
Density is the ratio of mass to volume (D = m/v). The standard for density is water, which at 4° Celsius has a density of one gram per milliliter. The density of all other substances is determined against this standard.

This activity may be used as an open-ended or fully-structured activity. The open-ended approach is strongly recommended for use in assessing the understanding of those students who have had sufficient experience with density investigations. The first student page is intended for this approach. The *Key Question* is posed, but students must design and conduct their own investigation. They are asked to record important elements of their plan for conducting the investigation and reporting the results, but are free to add such other documentation as they see fit. An abbreviated table of densities or a science reference book that contains a more complete table should be made available.

The second page outlines the step-by-step procedure to be followed in the fully-structured approach. It is designed for use with students who have had little experience with density investigations.

With either approach, students need to be alerted to the fact that careful measuring of mass and volume is required to determine the density with sufficient accuracy to be able to identify the metal.

Generally, BBs are copper colored that may cause students to assume that they are made of copper which has a density of 8.9 grams per milliliter (or cubic centimeter). In most cases the label on the container notes that these are steel BBs. The type of steel used in BBs has a density of about 7.85 grams per milliliters. If the investigation is carried out carefully, students should be able to rule out copper or brass as the metal used.

Pouring BBs into a graduated cylinder will not be a sufficient way to determine volume. Air space in the packing must be considered. When BBs are poured into the graduated cylinder and firmly packed by shaking, they are said to be "randomly" packed. When spheres are randomly packed they will occupy no more than 63.66% of the space within 100 milliliters. To simplify the process in this activity, water displacement is used to determine volume.

### Management

1. BBs can be purchased in stores such as K-Mart, Wal-Mart, or gun shops.
2. If the supply is limited, teams can be rotated through a single center with the necessary supplies for this investigation.
3. It is important to thoroughly dry the BBs after use to avoid rusting and discoloration!
4. An abbreviated table of densities or a science reference book that contains a more complete table should be made available to those designing their own investigation.
5. To prevent any students from slipping, warn them not to let any BBs fall on the floor.

The following approach is offered for those students ready for more independent exploration.

*Open-ended*: Pose the *Key Question*. Students must design and conduct their own investigation by recording the important elements of their plan for conducting the investigation and reporting the results. They are free to add such other documentation as they see fit.

### Procedure

1. Ask students how the composition of BBs could be determined if the container did not provide that information. [Compare the density of the BBs to the densities of various metals.]
2. Have the students put 200 mL of water into the graduated cylinder. Direct them to find and record the mass of the cylinder and water.

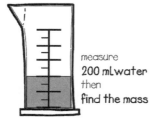

measure
200 mL water
then
find the mass

3. Have the students add BBs to the water until the water level is raised 50 mL. They should find and record the mass of the cylinder, water, and BBs. By subtracting the mass of the cylinder and water, they will find the mass of the BBs.

•Add BBs until the water level reads 250 mL

•Then find the mass

4. Have students add another 50 mL of BBs to the water, raising the level of the water in the cylinder to 300 mL. Direct them to find and record the mass of the cylinder, water, and BBs, subtract the mass of the cylinder and water to find the mass of the BBs.

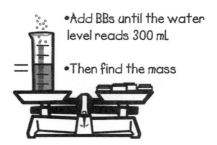

•Add BBs until the water level reads 300 mL

•Then find the mass

5. Once again, have students add another 50 mL of BBs to the water and find and record the masses as before.

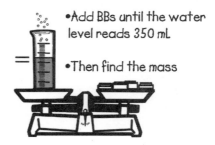

•Add BBs until the water level reads 350 mL

•Then find the mass

6. Direct them to find the density for 50 mL, 100 mL, and 150 mL of BBs.
7. Ask them to find the average density of the BBs and compare to the densities listed in the table to determine their composition.

## Connecting Learning

1. In what ways can the mass of BBs that fill a graduated cylinder to the 100 milliliter mark be determined? [By filling the cylinder to the 100 milliliter level and then by (1) pouring them into a pan of the balance and finding the mass directly or (2) using the indirect measurement approach outlined on the student page.]

2. In what other way could the density of the BBs be determined? [Crease a sheet of paper to create a trough. The trough will help align the BBs in a neat row. Place 20-50 BBs in the trough and manipulate until they form a neat row. Measure the total length of the row and count the number of BBs in it. Divide the length by the number of BBs to find the average diameter. In one instance, students found their BBs had an average diameter of 4.375 mm, and therefore, an average radius of 2.187 mm. From this the volume could be computed using the formula $V = 4/3\pi\, r^3$. Using 100 or more BBs (the larger the number the greater the accuracy), find their mass. Compute their total volume using your earlier data. Then divide the mass by the volume to obtain the density for comparison with the densities in the table.]

3. Why were you able to use the graduated cylinder and water to determine the volume of the BBs? [When an object is submerged, it displaces a volume of water equal to the volume of the object itself.]

4. What other objects could we use to determine their metallic composition?

\*   Reprinted with permission from *Principles and Standards for School Mathematics,* 2000 by the National Council of Teachers of Mathematics. All rights reserved.

# What's in a BB?

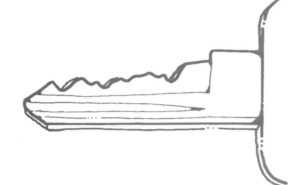

## Key Question

Of what substance are BBs composed?

## Learning Goal

### *Students will:*

find the density of BBs and match that against the density of metals to determine the substance of which they are made.

# What's in a BB?

## Of what substances are BBs composed?

To find the answer, determine the density of BBs and check your results with those in the table.

| Mass of 200 mL Water and Cylinder | Volume of BBs | Mass of Cylinder, Water, and BBs | Mass of BBs | Density of BBs (g/mL) |
|---|---|---|---|---|
| | 50 mL | | | |
| | 100 mL | | | |
| | 150 mL | | | |
| | Average Density | | | |

### Table of Densities

| | |
|---|---|
| Aluminum | 2.7 |
| Brass | 8.5 |
| Copper | 8.9 |
| Gold | 19.3 |
| Lead | 11.3 |
| Magnesium | 1.7 |
| Silver | 10.5 |
| Steel | 7.8 |

Judging by the density, it appears the metal is _____ .

Compare your results with other groups. The class average density is_____ .

# What's in a BB?

## Of what substance are BBs composed?

- Design and conduct your own investigation.
- Record your procedure, the data you gather, and the results of your investigation.

# What's in a BB?

## Connecting Learning

1. In what ways can the mass of BBs that fill a graduated cylinder to the 100 milliliter mark be determined?

2. In what other way could the density of the BBs be determined.

3. Why were you able to use the graduated cylinder and water to determine the volume of the BBs?

4. What other objects could we use to determine their metallic composition?

# The AIMS Program

AIMS is the acronym for "**A**ctivities **I**ntegrating **M**athematics and **S**cience." Such integration enriches learning and makes it meaningful and holistic. AIMS began as a project of Fresno Pacific University to integrate the study of mathematics and science in grades K-9, but has since expanded to include language arts, social studies, and other disciplines.

AIMS is a continuing program of the non-profit AIMS Education Foundation. It had its inception in a National Science Foundation funded program whose purpose was to explore the effectiveness of integrating mathematics and science. The project directors in cooperation with 80 elementary classroom teachers devoted two years to a thorough field-testing of the results and implications of integration.

The approach met with such positive results that the decision was made to launch a program to create instructional materials incorporating this concept. Despite the fact that thoughtful educators have long recommended an integrative approach, very little appropriate material was available in 1981 when the project began. A series of writing projects have ensued and today the AIMS Education Foundation is committed to continue the creation of new integrated activities on a permanent basis.

The AIMS program is funded through the sale of this developing series of books and proceeds from the Foundation's endowment. All net income from program and products flows into a trust fund administered by the AIMS Education Foundation. Use of these funds is restricted to support of research, development, and publication of new materials. Writers donate all their rights to the Foundation to support its on-going program. No royalties are paid to the writers.

The rationale for integration lies in the fact that science, mathematics, language arts, social studies, etc., are integrally interwoven in the real world from which it follows that they should be similarly treated in the classroom where we are preparing students to live in that world. Teachers who use the AIMS program give enthusiastic endorsement to the effectiveness of this approach.

Science encompasses the art of questioning, investigating, hypothesizing, discovering, and communicating. Mathematics is a language that provides clarity, objectivity, and understanding. The language arts provide us powerful tools of communication. Many of the major contemporary societal issues stem from advancements in science and must be studied in the context of the social sciences. Therefore, it is timely that all of us take seriously a more holistic mode of educating our students. This goal motivates all who are associated with the AIMS Program. We invite you to join us in this effort.

Meaningful integration of knowledge is a major recommendation coming from the nation's professional science and mathematics associations. The American Association for the Advancement of Science in *Science for All Americans* strongly recommends the integration of mathematics, science, and technology. The National Council of Teachers of Mathematics places strong emphasis on applications of mathematics such as are found in science investigations. AIMS is fully aligned with these recommendations.

Extensive field testing of AIMS investigations confirms these beneficial results.

1. Mathematics becomes more meaningful, hence more useful, when it is applied to situations that interest students.
2. The extent to which science is studied and understood is increased, with a significant economy of time, when mathematics and science are integrated.
3. There is improved quality of learning and retention, supporting the thesis that learning which is meaningful and relevant is more effective.
4. Motivation and involvement are increased dramatically as students investigate real-world situations and participate actively in the process.

We invite you to become part of this classroom teacher movement by using an integrated approach to learning and sharing any suggestions you may have. The AIMS Program welcomes you!

# AIMS Education Foundation Programs

## A Day with AIMS®

Intensive one-day workshops are offered to introduce educators to the philosophy and rationale of AIMS. Participants will discuss the methodology of AIMS and the strategies by which AIMS principles may be incorporated into curriculum. Each participant will take part in a variety of hands-on AIMS investigations to gain an understanding of such aspects as the scientific/mathematical content, classroom management, and connections with other curricular areas. *A Day with AIMS®* workshops may be offered anywhere in the United States. Necessary supplies and take-home materials are usually included in the enrollment fee.

## A Week with AIMS®

Throughout the nation, AIMS offers many one-week workshops each year, usually in the summer. Each workshop lasts five days and includes at least 30 hours of AIMS hands-on instruction. Participants are grouped according to the grade level(s) in which they are interested. Instructors are members of the AIMS Instructional Leadership Network. Supplies for the activities and a generous supply of take-home materials are included in the enrollment fee. Sites are selected on the basis of applications submitted by educational organizations. If chosen to host a workshop, the host agency agrees to provide specified facilities and cooperate in the promotion of the workshop. The AIMS Education Foundation supplies workshop materials as well as the travel, housing, and meals for instructors.

## AIMS One-Week Perspectives Workshops

Each summer, Fresno Pacific University offers AIMS one-week workshops on its campus in Fresno, California. AIMS Program Directors and highly qualified members of the AIMS National Leadership Network serve as instructors.

## The AIMS Instructional Leadership Program

This is an AIMS staff-development program seeking to prepare facilitators for leadership roles in science/math education in their home districts or regions. Upon successful completion of the program, trained facilitators may become members of the AIMS Instructional Leadership Network, qualified to conduct AIMS workshops, teach AIMS in-service courses for college credit, and serve as AIMS consultants. Intensive training is provided in mathematics, science, process and thinking skills, workshop management, and other relevant topics.

## College Credit and Grants

Those who participate in workshops may often qualify for college credit. If the workshop takes place on the campus of Fresno Pacific University, that institution may grant appropriate credit. If the workshop takes place off-campus, arrangements can sometimes be made for credit to be granted by another institution. In addition, the applicant's home school district is often willing to grant in-service or professional-development credit. Many educators who participate in AIMS workshops are recipients of various types of educational grants, either local or national. Nationally known foundations and funding agencies have long recognized the value of AIMS mathematics and science workshops to educators. The AIMS Education Foundation encourages educators interested in attending or hosting workshops to explore the possibilities suggested above. Although the Foundation strongly supports such interest, it reminds applicants that they have the primary responsibility for fulfilling *current* requirements.

**For current information regarding the programs described above, please complete the following:**

---

## *Information Request*

Please send current information on the items checked:

____ *Basic Information Packet* on AIMS materials
____ *AIMS Instructional Leadership Program*
____ *AIMS One-Week Perspectives* workshops

____ *A Week with AIMS®* workshops
____ Hosting information for *A Day with AIMS®* workshops
____ Hosting information for *A Week with AIMS®* workshops

Name _____    Phone _____

Address _____
      Street                      City                    State   Zip

---

**AIMS**
The Magazine

# *We invite you to subscribe to* AIMS®!

Each issue of AIMS® contains a variety of material useful to educators at all grade levels. Feature articles of lasting value deal with topics such as mathematical or science concepts, curriculum, assessment, the teaching of process skills, and historical background. Several of the latest AIMS math/science investigations are always included, along with their reproducible activity sheets. As needs direct and space allows, various issues contain news of current developments, such as workshop schedules, activities of the AIMS Instructional Leadership Network, and announcements of upcoming publications.

AIMS® is published monthly, August through May. Subscriptions are on an annual basis only. A subscription entered at any time will begin with the next issue, but will also include the previous issues of that volume. Readers have preferred this arrangement because articles and activities within an annual volume are often interrelated.

Please note that an AIMS® subscription automatically includes duplication rights for one school site for all issues included in the subscription. Many schools build cost-effective library resources with their subscriptions.

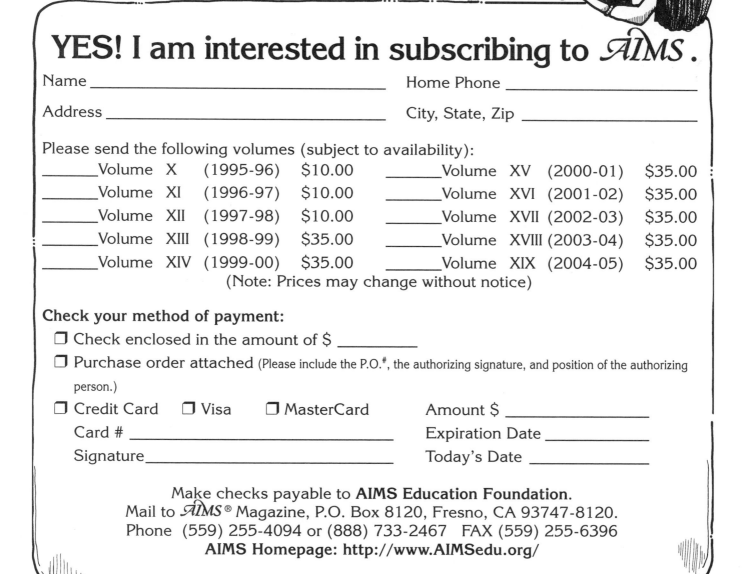

# YES! I am interested in subscribing to AIMS.

Name _____    Home Phone _____

Address _____    City, State, Zip _____

Please send the following volumes (subject to availability):

| _____Volume X (1995-96) $10.00 | _____Volume XV (2000-01) $35.00 |
| _____Volume XI (1996-97) $10.00 | _____Volume XVI (2001-02) $35.00 |
| _____Volume XII (1997-98) $10.00 | _____Volume XVII (2002-03) $35.00 |
| _____Volume XIII (1998-99) $35.00 | _____Volume XVIII (2003-04) $35.00 |
| _____Volume XIV (1999-00) $35.00 | _____Volume XIX (2004-05) $35.00 |

(Note: Prices may change without notice)

**Check your method of payment:**

❑ Check enclosed in the amount of $ _____

❑ Purchase order attached (Please include the P.O.#, the authorizing signature, and position of the authorizing person.)

❑ Credit Card    ❑ Visa    ❑ MasterCard    Amount $ _____

Card # _____    Expiration Date _____

Signature_____    Today's Date _____

Make checks payable to **AIMS Education Foundation**.
Mail to AIMS® Magazine, P.O. Box 8120, Fresno, CA 93747-8120.
Phone (559) 255-4094 or (888) 733-2467   FAX (559) 255-6396
**AIMS Homepage: http://www.AIMSedu.org/**

# AIMS Program Publications

Actions with Fractions 4-9
Awesome Addition and Super Subtraction 2-3
Bats Incredible! 2-4
Brick Layers 4-9
Brick Layers II 4-9
Chemistry Matters 4-7
Counting on Coins K-2
Cycles of Knowing and Growing 1-3
Crazy about Cotton Book 3-7
Critters K-6
Down to Earth 5-9
Electrical Connections 4-9
Exploring Environments Book K-6
Fabulous Fractions 3-6
Fall into Math and Science K-1
Field Detectives 3-6
Finding Your Bearings 4-9
Floaters and Sinkers 5-9
From Head to Toe 5-9
Fun with Foods 5-9
Glide into Winter with Math & Science K-1
Gravity Rules! Activity Book 5-12
Hardhatting in a Geo-World 3-5
It Must Be A Bird Pre-K-2
It's About Time K-2
Jaw Breakers and Heart Thumpers 3-5
Just for the Fun of It! 4-9
Looking at Geometry 6-9
Looking at Lines 6-9
Machine Shop 5-9
Magnificent Microworld Adventures 5-9
Marvelous Multiplication and Dazzling Division 4-5
Math + Science, A Solution 5-9
Mostly Magnets 2-8
Movie Math Mania 6-9
Multiplication the Algebra Way 4-8
Off The Wall Science 3-9
Our Wonderful World 5-9
Out of This World 4-8
Overhead and Underfoot 3-5

Paper Square Geometry:
    The Mathematics of Origami
Puzzle Play: 4-8
Pieces and Patterns 5-9
Popping With Power 3-5
Primarily Bears K-6
Primarily Earth K-3
Primarily Physics K-3
Primarily Plants K-3
Proportional Reasoning 6-9
Ray's Reflections 4-8
Sense-Able Science K-1
Soap Films and Bubbles 4-9
Spatial Visualization 4-9
Spills and Ripples 5-12
Spring into Math and Science K-1
The Amazing Circle 4-9
The Budding Botanist 3-6
The Sky's the Limit 5-9
Through the Eyes of the Explorers 5-9
Under Construction K-2
Water Precious Water 2-6
Weather Sense:
    Temperature, Air Pressure, and Wind 4-5
Weather Sense: Moisture 4-5
Winter Wonders K-2

## Spanish/English Editions*
Brinca de alegria hacia la Primavera con las
    Matemáticas y Ciencias K-1
Cáete de gusto hacia el Otoño con las
    Matemáticas y Ciencias K-1
Conexiones Eléctricas 4-9
El Botanista Principiante 3-6
Los Cinco Sentidos K-1
Ositos Nada Más K-6
Patine al Invierno con Matemáticas y Ciencias K-1
Piezas y Diseños 5-9
Primariamente Física K-3
Primariamente Plantas K-3
Principalmente Imanes 2-8

\* All Spanish/English Editions include student pages in Spanish
    and teacher and student pages in English.

## Spanish Edition
Constructores II: Ingeniería Creativa Con Construcciones LEGO® (4-9)
    The entire book is written in Spanish. English pages not included.

## Other Science and Math Publications
Historical Connections in Mathematics, Vol. I 5-9
Historical Connections in Mathematics, Vol. II 5-9
Historical Connections in Mathematics, Vol. III 5-9
Mathematicians are People, Too
Mathematicians are People, Too, Vol. II
Teaching Science with Everyday Things
What's Next, Volume 1, 4-12
What's Next, Volume 2, 4-12
What's Next, Volume 3, 4-12

For further information write to:
AIMS Education Foundation • P.O. Box 8120 • Fresno, California 93747-8120
www.AIMSedu.org/ • Fax 559•255•6396

# AIMS Duplication Rights Program

AIMS has received many requests from school districts for the purchase of unlimited duplication rights to AIMS materials. In response, the AIMS Education Foundation has formulated the program outlined below. There is a built-in flexibility which, we trust, will provide for those who use AIMS materials extensively to purchase such rights for either individual activities or entire books.

It is the goal of the AIMS Education Foundation to make its materials and programs available at reasonable cost. All income from the sale of publications and duplication rights is used to support AIMS programs; hence, strict adherence to regulations governing duplication is essential. Duplication of AIMS materials beyond limits set by copyright laws and those specified below is strictly forbidden.

## Limited Duplication Rights

Any purchaser of an AIMS book may make up to *200 copies* of any activity in that book for use at *one school site*. Beyond that, rights must be purchased according to the appropriate category.

## Unlimited Duplication Rights for Single Activities

An individual or school may purchase the right to make an unlimited number of copies of a single activity. The royalty is $5.00 per activity per school site.

Examples:  3 activities x 1 site  x $5.00 =  $15.00
9 activities x 3 sites x $5.00 = $135.00

## Unlimited Duplication Rights for Entire Books

A school or district may purchase the right to make an unlimited number of copies of a single, *specified* book. The royalty is $20.00 per book per school site. This is in addition to the cost of the book.

Examples: 5 books  x  1 site x $20.00 = $100.00
12 books x10 sites x $20.00 = $2400.00

## Magazine/Newsletter Duplication Rights

Those who purchase *AIMS®* (magazine)/*Newsletter* are hereby granted permission to make up to 200 copies of any portion of it, provided these copies will be used for educational purposes.

## Workshop Instructors' Duplication Rights

Workshop instructors may distribute to registered workshop participants a maximum of 100 copies of any article and/or 100 copies of no more than eight activities, provided these six conditions are met:

1. Since all AIMS activities are based upon the *AIMS Model of Mathematics* and the *AIMS Model of Learning*, leaders must include in their presentations an explanation of these two models.
2. Workshop instructors must relate the AIMS activities presented to these basic explanations of the AIMS philosophy of education.
3. The copyright notice must appear on all materials distributed.
4. Instructors must provide information enabling participants to order books and magazines from the Foundation.
5. Instructors must inform participants of their limited duplication rights as outlined below.
6. Only student pages may be duplicated.

Written permission must be obtained for duplication beyond the limits listed above. Additional royalty payments may be required.

## Workshop Participants' Rights

Those enrolled in workshops in which AIMS student activity sheets are distributed may duplicate a maximum of 35 copies or enough to use the lessons one time with one class, whichever is less. Beyond that, rights must be purchased according to the appropriate category.

## Application for Duplication Rights

The purchasing agency or individual must clearly specify the following:
1. Name, address, and telephone number
2. Titles of the books for Unlimited Duplication Rights contracts
3. Titles of activities for Unlimited Duplication Rights contracts
4. Names and addresses of school sites for which duplication rights are being purchased.

*NOTE: Books to be duplicated must be purchased separately and are not included in the contract for Unlimited Duplication Rights.*

The requested duplication rights are automatically authorized when proper payment is received, although a *Certificate of Duplication Rights* will be issued when the application is processed.

Address all correspondence to:  Contract Division
AIMS Education Foundation
P.O. Box 8120
Fresno, CA  93747-8120

www.AIMSedu.org/
Fax 559•255•6396